The Aspiring CIO and CISO

A career guide to developing leadership skills, knowledge, experience, and behavior

David J. Gee

‹packt›

The Aspiring CIO and CISO

Group Product Manager: Dhruv Jagdish Kataria

Publishing Product Manager: Dhruv Jagdish Kataria

Book Project Manager: Ashwini C

Senior Editor: Apramit Bhattacharya

Technical Editor: Arjun Varma

Copy Editor: Safis Editing

Proofreader: Apramit Bhattacharya

Indexer: Hemangini Bari

Production Designer: Vijay Kamble

DevRel Marketing Coordinator: Marylou De Mello

First published: June 2024

Production reference: 2160724

Published by Packt Publishing Ltd.

Grosvenor House

11 St Paul's Square

Birmingham

B3 1RB, UK

ISBN 978-1-83546-919-4

www.packtpub.com

This book is inspired by My wife Anna, the love of my life, my best friend, and the best teacher that I have ever had. My family - You will never know how proud I am of what incredible adults you have grown up to be. To my grandkids, Azalea (Azzy) and Harrison (Harry), who are the apples of my eye and whom I love dearly. Finally, of course, my parents David and Cindy, who have long passed this world, but whose positive role modeling I try to live up to everyday. (I've been a very blessed person.)

Foreword

For many IT professionals, the roles of **Chief Information Officer (CIO)** and **Chief Information Security Officer (CISO)** represent the pinnacle of their careers. The roles are challenging and demanding, and they offer the incumbents the opportunity to have a significant impact on the success of an organization. But how do you get there? What skills and experience do you need? How do you develop yourself to become a strong candidate for these coveted positions?

In this book, *The Aspiring CIO and CISO*, David shares his insights and experiences to help you navigate the path to becoming a CIO or CISO. David guides you through the critical aspects you need to consider.

My own career has been a wonderful journey of learning across different roles and companies. I didn't have any such reference that I could rely on and had to learn by myself. The book specifically talks about building a career as a CIO and CISO, and I would support that having this ability to pivot across both domains is a real career advantage.

This book is not just about getting that dream job; it's about building a fulfilling and successful career in IT leadership. Whether you're just starting out or looking to take your career to the next level, *The Aspiring CIO and CISO* provides the roadmap you need to achieve your goals.

I encourage you to take this journey and turn your aspirations into reality.

Darryl West

Former Global Group CIO, HSBC

Contributors

About the author

David J. Gee is a husband, father, and grandfather who just happens to have had the privilege of spending more than 25 years as a business leader in the roles of CIO, CISO, and technology, cyber, and data risk executive.

David has an eclectic background as a transformation change agent who has lived across five countries and worked in different industries, including banking, insurance, pharmaceuticals, building products, and media.

He won the Australia CIO of the Year in 2014 for a successful core, mobile, and online banking tranformation and the Global Leaders Award from FS ISAC in 2023 for his contributions to cybersecurity in financial services.

David has reinvented himself throughout his career and is now transforming into a non-executive director and board advisor. He is an avid writer and has published a few hundred articles for CIO, Computerworld, CSO (cyber), and ITnews. His articles have been translated into multiple languages.

As a venture capital partner, David has enjoyed connecting fintech firms with enterprises and helping these start-ups scale and grow.

About the reviewers

Sibylla Muecke is passionate about unlocking value for businesses through better decision-making. She is a lawyer with certifications in information management and leads initiatives in financial services across data, records and risk management, and regulatory policy and compliance.

Sibylla has received recognition and innovation awards throughout her domestic and international experience for contributions to business efficiency and building organizational capability.

Lily Couper is an emerging technology professional, currently specializing in technology, cyber, and data risk at Macquarie Group.

Lily is at the early stage of her career and pondering over longer-term career options, hence she has a strong personal interest in the subject matter of this book. She has a bachelor's degree in history as well as a bachelor of engineering degree with first-class honors from the University of Sydney.

Table of Contents

4

CIO and CISO Interview Tips   45

Part 2: What to Do in the First 90 Days

5

CIO – The First 90 Days   75

6

Part 3: Being the CIO or CISO

7

8

Understanding the Pressures CIOs and CISOs Face   147

9

CIO and CISO Survival Skills   165

Part 4: What's Next in Your Career?

10

Looking for the Next Elevator   185

11

12

Preface

Imagine that you are at the bottom of a mountain and making your way up the path. There is snow at the top of it and, along the way, many pointy rocks to navigate. Your destination is the summit, and there are many approaches that you can take to climb.

This book is intended to be your guide to reaching the summit of your career aspirations. I hope this book inspires the aspiring CIO and CISO to reach their career objectives. You can choose to walk up the mountain or take the gondola lift. The journey on the gondola lift will still have bumps, but you are able to traverse the distance safer and faster.

Being a CIO or CISO is an incredibly rewarding career journey. You will experience much personal growth and learning and face a new challenge to tackle every day. For many of you, that will be the key motivation for taking on this role, not the status, prestige, or rewards that may come from this position. It might be considered a personal test to see how much you can develop yourself.

There are many tricks and traps along the way in the career of a CIO and CISO. How can you prepare yourself for this journey? I recommend that you reflect on where you are and what you need to transform to make this a reality. I used the word *transform* on purpose as each leader will need to *stretch* into this new form.

Who this book is for

You are probably reading this as you would like to be a CIO or a CISO. Regardless of what stage you are at in your career – from starting out to being a senior manager – you might feel that there are gaps that you need to address to make this journey.

I've titled this book *The Aspiring CIO and CISO* on purpose as I have taken on both roles during the course of my career. Hence, I would encourage you to evaluate these opportunities equally. Both are worthy ambitions to pursue.

What this book covers

Chapter 1, *Starting the Journey to Become a CIO or CISO*, is the starting point of this journey. This chapter helps you to understand your current brand. Your brand is what qualities others associate with you. Your personal brand will dictate whether you are successful in becoming a CIO or CISO. The brand will shape your journey and prescribe what actions you need to take to address any of these perceived gaps. Thus, understanding what to refine and improve is a key factor.

Chapter 2, How to Develop Yourself to Be a CIO or CISO, explores the **Skills, Knowledge, Experience, and Behavior (SKEB)** that a CIO and CISO will require. There is a focus on soft skills that the CIO and CISO should aim to possess, and certain specific soft skills for these roles are essential. By the end of the chapter, you will know how to complete your own soft skills gap analysis and set some objectives to progress with these.

Chapter 3, Executing Your Career Path to Becoming a CIO or CISO, reviews how you can create your career and position objectives for your CV. The concepts of *stretch* and *becoming comfortable with being uncomfortable* are explored. We look at how to connect the dots on your career plan and try to think two jobs ahead, to ensure that you understand what SKEB you want to gain for this role to enable you to reach this position. I will introduce the concept of growing others to grow yourself. I also discuss different career path approaches that you may not have contemplated. Finally, we will review the CIO and CISO interview process.

Chapter 4, CIO and CISO Interview Tips, will delve into interview preparation to land your next CIO and CISO role. I outline the 25 most common questions that a CIO and CISO may be asked. Then I suggest 20 questions, which you should consider choosing two to three from, to ask the interview panel. By the end of the chapter, you will be ready to nail the interview.

Chapter 5, CIO – The First 90 Days, will show you how to build a plan for starting out as a CIO. I have included a template and described the work required to shape your own plan. There are working examples of how to engage stakeholders, review your IT strategy/roadmap, and engage your new team. I also talk about accelerating your own business learning and the key metrics that send a message to your team and key stakeholders. Then there is a retrospective review to see whether you need to update your 90-day plan for the next cycle. By the end of the chapter, you will be able to develop your own 90-day plan that is tailored to your new role as a CIO.

Chapter 6, CISO – The First 90 Days, will teach you how to develop your own 90-day plan for a CISO. There is a cyber strategy/roadmap to review and also stakeholders to engage. Once we have understood the stakeholder engagement mapping and plan for the CISO, we will work through an example. The new CISO has to orientate on key risk metrics, and some best practices are noted. There is a review of cyber governance processes, including frameworks to adopt. By the end of the chapter, you will be able to develop your own 90-day plan that is tailored to your new role as a CISO.

Chapter 7, Moments of Truth (When You Accelerate Your Growth), provides examples of when a CIO and CISO really take on their roles. These are moments that accelerate your learning and gain you respect from your key stakeholders and team. These are moments when you define yourself, and a few scenarios are explored to illustrate how this experience will reinforce positive behaviors.

Chapter 8, Understand the Pressures CIOs and CISOs Face, talks about the stress and pressure that is faced in a day in the life of the CIO and CISO. There are different types of CIO and CISO, and the stress indicators can vary dramatically based on the natural style that you bring to the table. Then, as a CIO, you have to work effectively with the CISO (and vice versa). Where you are both aligned and not aligned will have to be considered.

Chapter 9, CIO and CISO Survival Skills, explores Maslow's theory and how it applies to CIOs and CISOs. With this, detailed stakeholder analysis and approaches can be carried out and provide you with some valuable insights to manage these relationships. There is a discussion around building alliances and when to also look externally for mentors and coaches. Finally, we look at how to avoid workplace politics and ways to navigate certain difficult scenarios.

Chapter 10, Looking for the Next Elevator, deals with what you should do if you don't feel the role is a good fit. We will essentially evaluate what the right buttons to press are. There are times when a consulting gig makes sense before you consider returning to another CIO or CISO position. Taking a more holistic bird's-eye view and reflecting on your career will mean that you consider your life and career decisions closely coupled. Then, when you are ready to leave, we will explore how to efficiently hand over to your successor.

Chapter 11, Risk Management as a Career Option, is a bonus chapter in which I take you through a career path that you have probably never considered. I explore how your battle scars and SKEB have prepared you perfectly for this alternate career path. The chapter discusses a very different model of risk management than is typical, modeled on being a *coach* rather than a *player*, *referee*, or even *spectator*. By the end of this chapter, an alternative career door could have been opened.

Chapter 12, What CIOs and CISOs Do in Retirement, is the final chapter, where you will learn about the mountains you might want to climb next. We will explore some of the motivations you might have and the post-career moves that you can make. Again, given we want to always think two steps ahead, now that you are a CIO and CISO, you need to think about what is next. We will reflect on how to consider this to position yourself better for the future.

To get the most out of this book

As everyone will have a different starting point, you may want to read ahead to specific chapters depending on what is relevant to the position you are in. My guidance is that you start off by reading the first few chapters and then jump ahead to any chapters that are most relevant to you.

You will certainly have questions that you want to try to resolve, so in anticipation of this, I have made a note of 100 questions in the *Appendix* that you may have that this book can help you to try to answer. As you are working your way through this book, you may find that you want to make note of some additional questions that you would like to be answered. It is also up to you whether you want to satisfy your curiosity and jump to a chapter that answers a specific question, and not necessarily read this book from front to back.

Again, that's your choice, and your life and career are very much a journey of discovery. Each of us has to take this journey in a manner that works and makes sense.

Here are some key questions to ask yourself, numbered to correspond to the chapter in which they are answered:

1. Why do I need to build my own brand to be a CIO and CISO?

2. How do I develop my skills, knowledge, experience, and behavior to be a CIO or CISO?

3. How do I develop my career path to be a CIO and CISO?

4. How can I nail the interview for a CIO or CISO role?

5. How do I write my plan for the first 90 days as a CIO?

6. How do I write my plan for the first 90 days as a CISO?

7. How do moments of truth accelerate my growth?

8. How do I manage the stress that comes with the CIO and CISO roles?

9. What are the survival skills for a CIO and CISO?

10. How do I plan for my next CIO or CISO role?

11. Why should I consider Risk Management as a potential career path?

12. What do I plan to do in my retirement?

I'm sure that there will be many more questions that arise in your mind as you read this book. Indeed, I'm confident that you will encounter new questions to be addressed, and there are some areas where I won't be able to provide you with guidance.

Enjoy the journey and see you on the other side as you rise into your new role!

Get in touch

Feedback from our readers is always welcome.

General feedback: If you have questions about any aspect of this book, email us at customercare@ packtpub.com and mention the book title in the subject of your message.

Errata: Although we have taken every care to ensure the accuracy of our content, mistakes do happen. If you have found a mistake in this book, we would be grateful if you would report this to us. Please visit www.packtpub.com/support/errata and fill in the form.

Piracy: If you come across any illegal copies of our works in any form on the internet, we would be grateful if you would provide us with the location address or website name. Please contact us at copyright@packt.com with a link to the material.

If you are interested in becoming an author: If there is a topic that you have expertise in and you are interested in either writing or contributing to a book, please visit authors.packtpub.com.

Share Your Thoughts

Once you've read *The Aspiring CIO and CISO*, we'd love to hear your thoughts! Scan the QR code below to go straight to the Amazon review page for this book and share your feedback.

https://packt.link/r/1835469191

Your review is important to us and the tech community and will help us make sure we're delivering excellent quality content.

Download a free PDF copy of this book

Thanks for purchasing this book!

Do you like to read on the go but are unable to carry your print books everywhere?

Is your eBook purchase not compatible with the device of your choice?

Don't worry, now with every Packt book you get a DRM-free PDF version of that book at no cost.

Read anywhere, any place, on any device. Search, copy, and paste code from your favorite technical books directly into your application.

The perks don't stop there, you can get exclusive access to discounts, newsletters, and great free content in your inbox daily

Follow these simple steps to get the benefits:

1. Scan the QR code or visit the link below

https://packt.link/free-ebook/9781835469194

2. Submit your proof of purchase
3. That's it! We'll send your free PDF and other benefits to your email directly

Part 1:
Your Journey to Becoming a CIO or CISO

In this first part, you will get an overview of the role of the CIO and CISO and will start mapping your own personal journey to this destination. We will cover the development of your brand and your overall gaps across **skills, knowledge, experience, and behavior (SKEB)**. We will talk about how to reflect on your soft skills and go outside of your comfort zone to grow as you prepare for the role. We will explore how helping others develop will help you get ready faster and be successful in a CIO or CISO position. Getting the CIO or CISO role will be challenging, so interview preparation is key. We will look at the questions that might be asked in the interview and then how best to ask probing questions yourself.

This part has the following chapters:

- *Chapter 1, Starting the Journey to Become a CIO or CISO*
- *Chapter 2, How to Develop Yourself to Be a CIO or CISO*
- *Chapter 3, Executing Your Career Path to Becoming a CIO or CISO*
- *Chapter 4, CIO and CISO Interview Tips*

1

Starting the Journey to Become a CIO or CISO

This book is for leaders who aspire to be a **Chief Information Officer (CIO)** or **Chief Information Security Officer (CISO)** and provides practical guidance as to how to build a career as a CISO or CIO. You've likely opened this book as you have a desire to achieve one of these senior positions. There are few more challenging and interesting roles than these. I've written this book as a guide to a younger version of myself, when I was filled with more questions than answers and some degree of uncertainty about the direction of my career.

> The fundamental question that is to be addressed in this chapter is as follows:
>
> Why do I need to build my own brand to be a CIO and CISO?

Although the career path to be a CIO is traditionally different from that of a CISO, this book could also be for leaders who aspire to be a CIO *and* CISO. Many IT professionals would naturally not consider both as an option and will instead choose only one of these tracks. This book will provide insights into tackling either or both career paths. You might want two shots on goal or one – either would lead to a great and rewarding career. However, having a choice is a powerful option to consider.

In this chapter, we'll gain a deep understanding of what these roles entail and touch on the typical career paths that have taken many on the journey to become a CIO and CISO. Then, we'll look at defining your personal brand and what you may need to change to achieve either of these career paths. This is because having a good understanding of your personal brand is a critical step on your path to becoming a CIO or CISO. Should you embark on this journey without a clear understanding of your brand, then it will only make it more difficult.

In this chapter, we will cover the following main topics:

- Understanding the CIO and CISO roles
- Introducing the CIO career path
- Introducing the CISO career path
- What is your current brand?
- What do you need to change?

Understanding the CIO and CISO roles

You likely already have a fundamental understanding of these roles. It is also likely that you have worked for a CIO or CISO or directly reported to them, and that you have made some personal observations about what the roles entail. We should, however, note that no two CIO and CISO roles are exactly the same, and there will be some differences in their responsibilities. First, let's look at the role of the CIO and what it requires.

The role of the CIO

The role of the CIO has changed dramatically over time. It has evolved over many years, with different titles and responsibilities, as IT itself has grown and developed. Some examples of former titles of the role are as follows:

- Information Services Manager
- IT Director
- MIS Manager

Despite the variation in the title, the CIO has always been principally responsible for managing IT infrastructure and operations. This role of overseeing technology has evolved from centralized mainframe systems to distributed computing, and now, cloud-based processing.

The CIO has seen their responsibilities *morph* with tasks added and removed. This was caused by the role becoming too large for one individual and by the expansion of the skills required. As these changes occurred, there were new C-suite colleagues added to some organizations, and these include the following:

- Chief Digital Officer
- Chief Data Officer
- Chief Knowledge Officer

- Chief Technology Officer
- Chief Transformation Officer
- Chief Innovation Officer

In large, complex enterprises, the CIO may cover all or some of these roles. We should expect that this rate of change will continue, given that technology continues to strategically drive change in organizations.

In a classic sense, the specific responsibilities of a CIO usually include the following:

- Taking ownership of developing and executing the IT strategy
- Effectively working with business stakeholder leaders to drive change through new technology
- Overseeing the development and system maintenance of IT applications, systems, and technology infrastructure
- Managing the IT portfolio, including the budget and resources
- Managing the security of the enterprise's data and networks

It's clear that the role of the CIO is more important than ever. CIOs are responsible for ensuring that their organizations have the technology they need to be competitive and successful. Let's look at some key points in terms of how this role has changed over time.

Traditional CIO	Today's CIO
Run legacy business	Design new digital business
Automate processes	Transform business processes
Conduct internal operations	Engage in design thinking (customer experience)
Oversee centralized IT	Implement distributed cloud model

Table 1.1 – Changes in the focus of the CIO role

There are CIOs that would be a mix of these two extremes, as well as individuals that operate at each end of this spectrum. Next, let's look at how the CISO role has evolved.

The role of the CISO

In contrast to the CIO role, the role of the CISO is relatively new but has also evolved. The position was previously called **Information Security Manager**, **Vice President Information Security**, or some other similar variation. However, the role is not the same – as cybersecurity became more strategic, this role became elevated in seniority. The CISO's role is simply to defend the enterprise data, systems, and assets from cyber threats. The CISO is the most senior-level executive responsible for protecting the organization, and their role usually includes the following:

- Managing the response to security incidents
- Overseeing security assessments of new systems and partners
- Engaging in people management of the cyber team
- Engaging in operational management of cyber technology
- Taking ownership of the cyber risk culture and staff education

The following table outlines the changes in this role over time:

Information Security Manager	Today's CISO
Manage security operations	Develop a cyber strategy
Implement security projects	Deliver a cyber transformation program
Manage unauthorized access to data	Manage enterprise risk and posture
Oversee developer access	Drive DevSecOps

Table 1.2 – Changes in the focus of the CISO role

As cybersecurity has become more critical for enterprises, the role of the CISO has evolved into that of a strategic leader who works with the board and management. The CISO can also be asked to interface externally with regulators, customers, and partners.

Now let's look at the typical career trajectories that can be followed to become a CIO.

Introducing the CIO career path

The CIO career path often follows the route of being promoted for strong delivery, particularly of evident transformation efforts acknowledged by the business executives. As such, there is a stronger weighting of CIOs who have a *program and project delivery* background than *operational experience*. As cyber-attacks spread across all sectors, we see that the CIO must pay more attention to cybersecurity. The CIO can only ignore cybersecurity at their own peril, as the stakes are high, and reputations can be severely impacted.

With that said, there are many paths to becoming a CIO. Here is an example of a typical CIO career path:

1. Commence your career as an **IT analyst** or **Support analyst**. This will provide you with a foundation in IT skills, knowledge, and experience of working across the IT portfolio and with different aspects of the business.

2. Take opportunities to build your skills, knowledge, and experience in **development**, **network admin**, or **project management**. These assignments will provide chances to build proficiency.

3. Either through internal promotion or external opportunity, you will take your first IT management position – **IT manager** or **IT director**. This is your first step into management and provides a more extensive role scope with larger teams. More importantly, you have the responsibility to own and implement the CIO's IT strategies.

4. It is likely that you will perform several iterations of *step 3* across different teams and with increased responsibility. During *steps 3 and 4*, there is normally an expectation that you understand the business and not just technology.

5. Promotion to **Chief Technology Officer** or **Transformation Officer** would be your first step into senior IT management. You become responsible for developing and executing the enterprise IT strategy and start to take on senior business stakeholder engagement.

6. Finally, we come to the **CIO** role – this may be an internal or external opportunity. You will be in charge, and the buck stops with you. Your role reports to the CEO or COO, and there are much higher expectations of you to deliver on the enterprise business objectives.

This example is very close to my own experience of becoming a CIO. There are more roads to becoming a CIO, and what is key is developing a brand that your stakeholders will associate with their idea of what a CIO is. Once you reach the CIO role, then the journey is not over and there are more senior CIO roles – Group CIO, CIO in larger enterprises, industries, and companies with more complex challenges to tackle.

Next, let's look at the CISO career path.

Introducing the CISO career path

The role of the CISO, as I have mentioned, is newer than the CIO role and is thought to have first appeared late in the 1990s. There is much more variation in how someone gets to become a CISO. In my case, my path shifted from that of a CIO to a CISO path after a long CIO career, managing CISOs and Information Security Managers for many years. However, as with the case of CIO, there are many paths to becoming a CISO as well. Here is an example of a typical CISO career path:

1. Commence your career as an **IT analyst** or **Support analyst**. This will provide you with a foundation in technology skills, knowledge, and experience of working across IT and different aspects of the business.

2.   Take opportunities to build your skills, knowledge, and experience in **security**, **network admin**, or **cloud project management**. These assignments will provide chances to build your technology proficiency.

3.   Through internal promotion, take on more expert roles – for instance, **incident response**, **penetration tester**, **forensics analyst**, **cyber intelligence**, **security consultant**, or **architect**.

4.   Advance to a **security manager** position. This will give you experience in leading and managing teams and implementing security programs. There may also be security director roles that you can take on that have greater responsibility.

5.   Finally, we come to promotion to the CISO role. CISOs are responsible for the organization's overall cyber strategy. The role typically reports to the CIO. However, it can also be aligned with the COO, given the critical nature of the position. The CISO is expected to understand the business and will also be expected to have strong stakeholder management skills.

The career path of the CISO has also been influenced by the acute shortage of staff with a security background. Hence, staff that have adjacent skills, knowledge, experience, and behavior, such as people experienced in network and cloud, can quite easily transition to this career path. There is also a well-trodden path of staff that have come from defense backgrounds, with operational intelligence experience, that have forked onto the CISO career path.

Now that you're familiar with the CIO and CISO roles, let's look at how your personal brand can help you on this chosen career path.

What is your current brand?

Your **personal brand** may or may not be clear to you. Your current brand can help or hinder the career ambitions that you hold. What I always have believed to be the litmus test for your personal brand is to find out the following: *What do people say about me when I am not in the room?* Like it or not, that is your personal brand.

There is no right or wrong brand. There are different types of CIOs and large variations in CISOs. However, your brand will be a constraining force that could cause friction against your ambitions. Do some analysis by writing down on a blank sheet of paper the attributes that you think others would say you have. These will be, by nature, both *positive* and – dare I say – *negative* traits. The following are examples of some positive traits of mine:

```
┌─────────────────────────────────────────────────┐
│                                                   │
│              David's current brand                │
│                                                   │
├───────────────────────────────────────────────────┤
│                                                   │
│    +   Strategic thinker                          │
│                                                   │
│    +   High energy                                │
│                                                   │
│    +   Seasoned tech and cyber executive          │
│                                                   │
│    +   Direct and transparent approach            │
│                                                   │
│    +   Confident and intuitive                    │
│                                                   │
└─────────────────────────────────────────────────┘
```

Figure 1.1 – David's brand

Once you have drafted what you believe are your qualities, this can be tested by asking your colleagues and supervisor for their inputs. My advice is to not show them your draft – instead, let them know you are reflecting on career ambitions and want to get a baseline of your personal brand. Your partner at home can usually also easily provide this feedback. Finding someone that you trust who has your best interests at heart but who is also willing to be honest and provide you with these insights is critical.

Remember your brand is what people say about you when you are not there, so this may be quite difficult, and perhaps also confrontational, for the person that you have requested this input from. Indeed, it could also be difficult for you.

Taking this example, the question is *What parts of my personal brand can I change?* It may not be possible to change traits, so this will require deep reflection and thought.

To be a CISO or CIO, what do you need to change?

You will have been exposed to one or many different CIOs and CISOs in your career. They too will have a personal brand and you will have taken note of their attributes, and will also have noted which of these are the weaker and stronger traits. These CIOs and CISOs may either be ones that you have worked for or observed in broader industry events and collaborations. I'd say that the larger the sample set, the better your sense of what good looks like – and, for that matter, what bad looks like. Both ends of the spectrum can give you traits to copy and avoid.

Taking my own example, in my opinion, it is fitting to first focus on strengthening my positive traits rather than trying to address my weaker points. For most of my career, I've been told that **strategic thinking** is one natural strength that I possess. While I'm not sure when this became part of my brand, I would note that there is some element of this being innate – meaning I was born this way. However, I've also learned that this is what I care about; I have a strong personal interest in sharpening my thinking and honing this strength. Being a strategic thinker is a very good attribute that a CIO or CISO should possess. However, it may not be what you have documented in your assessment, and, indeed, not all in the CIO/CISO C-suite have this trait as a strength.

In my case, I've learned that what is behind this drive to continually *sharpen my sword* is a personal insecurity that I will become unessential and outmoded in this world of change. For a person who has made a career of delivering transformational change, this may sound a little silly, but the key is that I try to understand why I ruminate over certain things and exploit this for my own personal development.

As a CIO or CISO, one of your key responsibilities would be to develop an **IT and cyber strategy**. A strategy is simply deciding what to do and what not to do. In practice, it is not that easy, and the more that you do this, typically, the better you will become at mastering it. How you *practice* this will depend on where you are in your current career – we will cover in the next chapter how to develop these action plans.

But let's continue with one of my examples – **high energy**. For me, this is a natural trait. I get energy from discussing ideas with others and making progress. In any normal meeting, I will be noticed for being habitually *active* and asking the right questions at the appropriate time. I use these interactions to learn about topics and understand the facts. Using my **confident and intuitive** nature, I sense the direction in which things should move and don't lack the courage to test the positions of others.

A CIO or CISO doesn't need to be a high-energy individual – in fact, most of the people that I have met are not. Many are reserved and analytical. To me, the secret is answering the following questions:

- How do you make what you have work for your brand?
- If you have traits that could be career derailers for the role that you aspire to, how can you develop a plan to make them less of a weakness?

Your personal brand will be what management or a recruiter will consider when you are in the selection process for a CIO or CISO role. They will review your CV and experience, accompanied by a reference check. The interview processes for these senior roles are usually much more robust than any other process that you will have encountered.

I once secured a CIO role that required 15 face-to-face interviews, via video conferencing – with one of them being held at midnight to meet the Global CIO of the enterprise. Some of those types of interviews are for *selection* and many others are to check your *social fit* with colleagues, peers, and – at times – even direct reports. There will be different selection methodologies applied and questions to probe strengths and weaknesses.

This will often involve probing weaknesses from the psychometric tests you might be required to complete. Moreover, the interview panel always looks at team fit, and whether they think your style will work in the new role. As CIO and CISO roles are considered strategic for the enterprise, they will want to understand your personal brand and how your leadership will drive the organization forward.

The reference checks for senior roles are particularly robust and there will be questions around how the candidate operates under pressure and how they are valued in their current role. If there are signals that the candidate's personal brand is not consistent with what is being portrayed, it could signal the end of the process for the candidate. Being trustworthy and trusted is a critical part of your brand if you ever want to be a CIO or a CISO. Moreover, it is important to be who you are. That said, everyone, including me and you, has areas to develop and improve. This is the most critical advice that I can give you at any stage of your career when you aspire to be a CIO or CISO.

The following figure is an example that highlights my brand and attributes that colleagues have provided feedback to me on:

David's current brand

+   Strategic thinker

+   High energy

+   Seasoned tech and cyber executive

+   Direct and transparent approach

+   Confident and intuitive

-   Lack focus on details

-   May not finish everything I start

-   Don't bring others with me on the journey

Figure 1.2 – David's brand as perceived by others

Seeking feedback from a broad range of people and getting honest input is important. I have learned that strengths can be good attributes, but if they're overused, then they detract from my brand.

You will also notice that there are more positive than negative feedback points. This is a natural part of the process – when I provide feedback, it is always important to have balance, and this does not mean equal numbers. Humans have a natural tendency to over-focus on negative feedback. Hence, we should always attempt to offer twice as many positive as negative points.

In my example, for all three negative traits, I have made a conscious effort to develop in those respects and make the traits less obvious. If I re-did my assessment today, I would certainly strike off two of the three as being part of my brand. For instance, *Bringing others with me* and *Attention to detail* can now be considered as strengths of mine. You may be asking how I addressed this gap. It has taken years of focus and concerted effort to do the things that I now do well. There is no secret – just be aware of the perceived gap and reflect on it as you take action in your role. If you have a proactive and positive view of bridging the gap, then it can be addressed. Of course, you should take feedback from your team and stakeholders – that is a simple way to conduct a pulse check on these areas.

Summary

The path to becoming a CIO or CISO is not an easy one. There will be numerous obstacles on your path when climbing that mountain. The route can be slippery with some jagged rocks, and as you ascend, there will be unwelcoming, cold conditions to greet you. But this is a journey and challenge that you want to take on.

In this chapter, you were introduced to the potential career trajectories of a CIO and CISO. This should help you have a better idea of where you currently are on your journey. You were then introduced to the idea of a personal brand. This can essentially be considered the starting point of your journey to understand your current brand. What do others say about you when you're not in the room? These are your strengths and areas for improvement. There will always be *work-on* items. The real question is whether you have truly embraced this or just begrudgingly listened when this feedback was provided. Your personal brand will dictate whether you can achieve a CIO or CISO role. Your brand will shape your journey and prescribe what actions you need to take to address any perceived gaps.

The next chapter looks at an approach to developing yourself according to your own personal career goals.

2

How to Develop Yourself to Be a CIO or CISO

Some of you may aspire to be a CIO or CISO. The goal of this chapter is to provide insights that will help you progress in your career to become a CIO or CISO. In many ways, this will give you *two shots on goal* for two very senior and strategic roles. There are also other roles adjacent to that of the CIO: chief technology officer and chief digital officer. This chapter will provide insights that are applicable to all of these roles as many of these roles are interchangeable in terms of the attributes required.

> **The fundamental question that is to be addressed in this chapter as follows:**
> How do I develop my skills, knowledge, experience, and behavior to be a CIO or CISO?

To successfully achieve your career goal, you will embark on a journey that will require you to change and grow. There are very few individuals who are ready from the get-go to be a CIO or CISO.

Therefore, the question is where do you invest the most effort to get this payback? Every road will take you to a destination, and each path can be rewarding and fulfilling; it just depends on what you seek.

In this chapter, we will cover the following topics:

- Building your development plan – the **Skills, Knowledge, Experience, and Behavior (SKEB)** model
- Why do soft skills matter?
- Understanding the gaps in your soft skills

Building your development plan – The SKEB model

Many years ago, in one of my expatriate CIO assignments, there was a strong focus on developing the middle management of a large organization. The CEO, who was my manager, had a dilemma: there were many senior executives who performed well but did not develop their direct reports effectively. For many years, they had been recruiting strong MBAs or other reputable graduate degrees to enhance the skill level of the team; however, there was a fundamental issue – the professional development plans of the teams were weak. This resulted in slow development, and often, staff with high-potential left for greener pastures.

The CEO decided to have a senior manager, who would develop a good development plan framework and act as a role model for the rest of the team. Then they would teach the middle management in all-day training sessions, rather than using external specialist trainers or consultants. This would send a message to staff that the exercise was critical and had to be taken seriously. Each of the executives was taught a simple but powerful approach, the **SKEB** model, that was easy to remember and action with their teams. The SKEB competency model is a framework first developed for evaluating the skills, knowledge, experience, and behavior of individuals in the construction industry. This model has now been applied more broadly. It is used to assess an individual's ability to perform their job effectively and safely. It can be represented as follows:

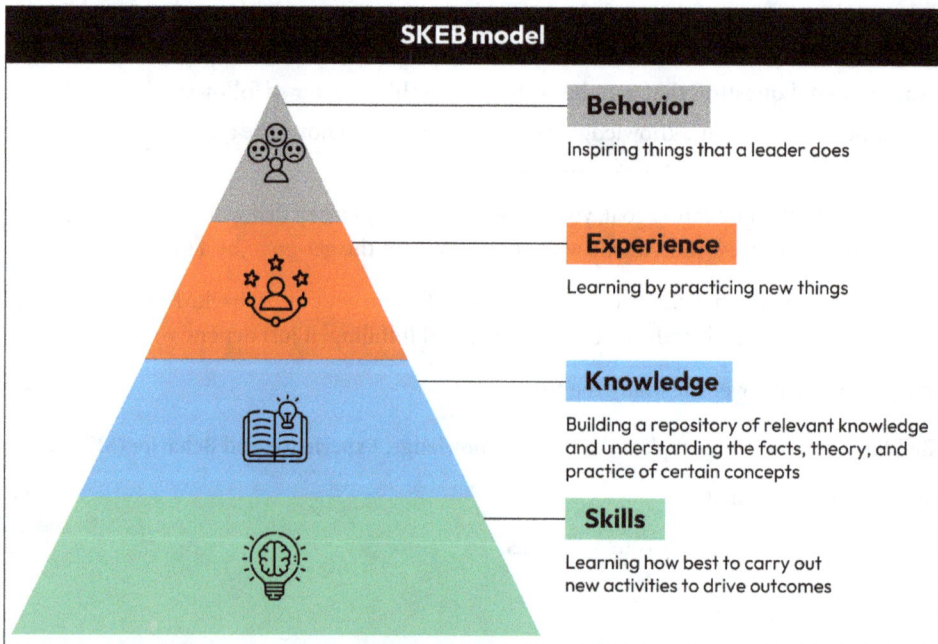

SKEB model

Behavior
Inspiring things that a leader does

Experience
Learning by practicing new things

Knowledge
Building a repository of relevant knowledge and understanding the facts, theory, and practice of certain concepts

Skills
Learning how best to carry out new activities to drive outcomes

Figure 2.1 – What is SKEB?

The model is simple but powerful, and I've applied it across many industries and large enterprises. When you are young and want to grow in your career, your focus is naturally on the development of *skills* and *knowledge*. These are foundational elements that can be transferred from role to role and will be used throughout your career. There are some skills, such as database management, programming, and project management, that are immediately transferable between companies. Early in my own career, I learned to develop these skills, which became a solid foundation that I still use every day in my roles as a CIO and CISO.

Some examples of skills that I have been able to apply across industries, companies, and countries are as follows:

- Problem-solving
- Software coding and testing
- Data modeling and analytics
- Project management
- Operations management (troubleshooting support etc.)
- Web and mobile development
- Network, cloud, and server administration
- Six Sigma analysis
- Agile development
- Report writing

The key when starting out in your career is to seek out roles that enable further development of these types of skills. When you become a CIO and CISO, there will be less time and opportunity to develop these skills and your focus will be on experience and behavior. These are a combination of business and technology skills, together with soft skills, which we will discuss later in this chapter.

The SKEB model is simple but requires a little work to make it relevant to you. Let's explore a few examples of the SKEB model to bring this to life:

SKEB competency	Examples
Skills: Your ability to perform specific tasks or activities	<ul><li>Data- and process-modeling skills</li><li>Coding (Python, C++, Java, PHP, JavaScript, etc.)</li><li>Managing OS environments (Kali Linux etc.)</li><li>AI, ML, and data analytics skills</li></ul>

SKEB competency	Examples
Knowledge: Your understanding of tech, data, and cyber concepts	<ul><li>AWS, GCP, Azure cloud certifications, and network design (routers, firewalls, VPNs, etc.)</li><li>**Large Language Model (LLM)** Generative AI fundamentals</li><li>Understanding of banking business processes</li></ul>
Experience: Your demonstrated ability to apply skills and knowledge in real-world situations	<ul><li>Risk and audit analysis</li><li>Security incident handling and response</li><li>Regulatory compliance</li><li>Experience in managing remote staff</li></ul>
Behavior: Your ability to demonstrate the values and behaviors expected of a CIO and CISO	<ul><li>Delivering a strategy that is complex and engages difficult stakeholders</li><li>Demonstrating strategic leadership across business units, projects, and operations</li><li>Being a role model for technology engagement with business</li><li>Having a track record of exceptional delivery</li></ul>

Table 2.1 – SKEB competency examples

As an example, we can gain *knowledge* about LLMs that we apply in the workplace, which can develop into the LLM *skill*, which will over time allow us to gain *experience* in driving an enterprise LLM program of work. Finally, to be successful at what we do, we will need to demonstrate leadership *behavior*.

A suggestion I would make here is to imagine what the required SKEB would be for a role that is two steps above yours. This is discussed further in the next chapter – but a summary is that reflecting on what is required in terms of skills, knowledge, experience, and behavior for that role two steps above will allow you to incrementally begin preparing yourself for entering that new territory.

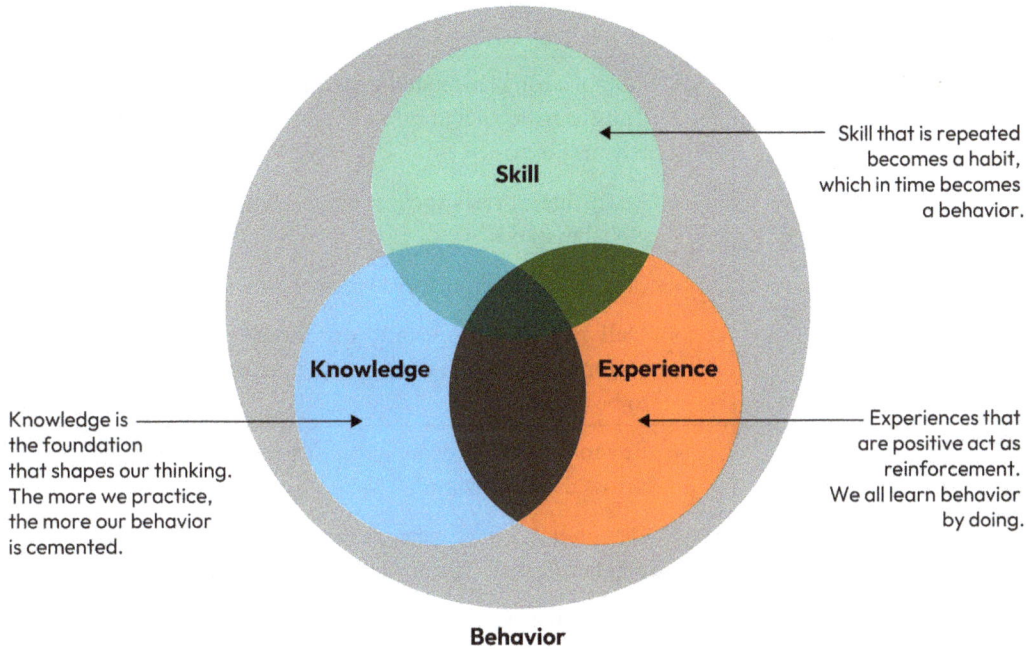

Figure 2.2 – The SKEB model in action

As you grow and develop into more senior roles, the *experience* and *behavior* become more important to focus on. These are harder to develop as they require specific projects or opportunities to build and practice. Each person who aspires to be a CIO or CISO will need to identify the **stretch development goals** that will provide them a platform to develop into that role. Even if you are a CIO or CISO, you will want to continuously stretch your learning goals. Here is a personal example of a plan that I prepared for myself in the past:

David's SKEB competency	Examples
Skills	<ul><li>Undertake a Black Hat conference hands-on hacking training course to further develop my technical cybersecurity skills</li><li>Improve my control management skills by growing my understanding of how to design better key control indicators</li></ul>
Knowledge	<ul><li>Deepen knowledge of the business to be able to understand digital and cyber connections</li><li>Grow my cloud security knowledge by undertaking professional training (e.g., CSSP) and grow the knowledge of my team</li><li>Build knowledge of defense-in-depth of XYZ enterprise</li><li>Develop my knowledge base around AI and LLMs by reading broadly and networking with external industry colleagues</li></ul>

David's SKEB competency	Examples
Experience	• Obtain strong leadership experience to drive cybersecurity transformation adoption across the different geographic and business units • Be the enterprise's spokesperson for regulators in the key global markets • Build extensive experience to present cybersecurity to the board • Deliver cyber training to our bank's external strategic commercial banking customers to enable a trusted brand to be built
Behavior	• Be a role model to encourage a cyber risk culture of ownership across the organization • Demonstrate a track record of exceptional delivery to remediate regulatory projects from key regulators • Be able to drive strong adoption of DevSecOps across digital and cyber teams to reduce the risks from code development

Table 2.2 – David's SKEB example

There are a few key observations that can be made from this SKEB example.

Old dogs *can* learn new tricks. There are always gaps in what you know or knowledge that can be enhanced. It does not matter who or how senior you are, there is always room to grow and learn. So, please think broadly about what you need to do differently to be more effective.

You may notice that the lists of *Experience* and *Behavior* items outnumber those of both *Skills* and *Knowledge*. This is partially due to seniority, which would be typical considering my experience. However, I would note that it is not merely about quantity; the elements that are noted for *Experience* and *Behavior* are and should be more difficult and therefore require more effort.

These are *experiential* elements that are learned in the role, and in your case, it is about practicing before you become a CIO or CISO. Remember, when you act and dress in a certain way, you are halfway to achieving your objective.

Figure 2.3 – Dress for success

While you may have identified some new objectives in your own SKEB analysis, soft skills will help drive your growth. Let's understand the importance of this in the next section.

Soft skills are hard – Why do they matter?

It is true that soft skills are difficult to acquire; otherwise, we would all have them already. But you must also work on specific soft skills that are required for the CIO and CISO roles. For many of us, the gaps in our soft skills will require significant attention. This is not something that can be fixed overnight; it requires life-long learning.

But which soft skills matter? Well, both the CIO and CISO are senior-level executives who require a range of soft skills to effectively lead and influence within their organizations.

The CIO must oversee the IT strategy and technology departments. Sometimes they must also manage digital transformation efforts and other teams within operations and procurement. However, the CISO is responsible for the cybersecurity strategy and managing the defense of the enterprise.

At times, the CIO and CISO will be in a relationship where one has to report to the other, and there will be tension around conflicting priorities and expectations. A number of these soft skills are required across both roles and should be considered *shared* soft skills. I personally refer to these as *superpowers* – if you have these, they will accelerate your growth to becoming a CIO or CISO.

Figure 2.4 – Soft skills are superpowers for the CIO and CISO

Let's discuss ten *shared* soft skills that are ideal for those aspiring to be a CIO or CISO. While both the CIO and CISO share some soft skills, their specific responsibilities lead to differences in the weighting of the skills' importance. Both roles require strong leadership, communication, and adaptability skills to be effective. CIOs will need to focus more on business alignment, innovation, and technology management, while CISOs must pay more attention to cyber, risk management, and incident response.

You will find that you have some of these skills and others will require your focus:

Soft skills	CIO	CISO
Leadership	The CIO needs strong leadership skills to guide their IT teams and align technology initiatives with the overall enterprise goals. They should inspire and motivate their teams to achieve strategic objectives.	The CISO will also need strong leadership skills to guide the cyber team and obtain the support of other executives and employees. The CISO must set the vision, inspire others, and make tough decisions.

Soft skills	CIO	CISO
Strategic thinking	The CIO must have a strategic mindset as they will need to develop and execute IT strategies that drive business change. They will need to get alignment of their IT initiatives with the longer-term business vision.	The CISO must develop and implement a cybersecurity strategy and roadmap aligned with the business and the CIO. There may be tension with the CIO over budget, as the cyber investment may eat into the budget for digital change.
Communication	Effective communication is critical for the CIO. They must convey complex tech concepts to business stakeholders. The CIO must communicate the value of IT investments and encourage collaboration between business units.	Effective communication is also critical for the CISO. The realm of cybersecurity is full of technical jargon, and the risk is not always well understood. The CISO must communicate effectively to build strong relationships with business unit departments and IT colleagues to get things done.
Problem management	The CIO encounters complex technology challenges daily. Having good problem-management skills is mandatory for success.	Cyber challenges are extremely complex and constantly changing. You are unlikely to encounter the same zero-day threats in two different enterprises. The CISO must have strong problem-solving skills to address new threats.
Risk management	The CIO is responsible for assessing and managing risks related to IT programs, projects, and cybersecurity. This creates a natural tension as resources must be assigned to these change projects or for business as usual. The CIO will make risk management decisions with input from the CISO.	The CISO performs risk management daily. This includes assessing cybersecurity risks, prioritizing them, and then carrying out risk mitigation. The CISO must also be able to explain risk management concepts to peer executives and the board so that they can understand it.

Soft skills	CIO	CISO
Adaptability	The CIO operates in an industry where change is constant. They should be open to learning and embracing change as there is a need to adapt to new trends, tools, and methodologies.	The CISO must be able to operate within a fast-paced environment. The landscape is changing and the CISO must try to stay ahead of cyber threat actors.
Emotional intelligence	The CIO must build strong working relationships and be able to work through the various mazes that exist. There is a strong need for emotional intelligence to navigate these corridors successfully.	The CISO must be able to deal with staff being under pressure and stress. This includes having a good understanding of and being able to manage emotions, as incidents can be tiring and lead to significant conflicts between teams and staff. The CISO must also be able to deal with their own work stress, requiring emotional intelligence.
Time management	CIOs work on multiple projects simultaneously. There is never enough time to do everything, so effective time management will help them prioritize and meet strategic and operational deadlines.	The CISO must also manage multiple moving parts – cyber projects, incidents, and other governance responsibilities. Effective time management will help keep things on track.
Team building	The CIO must be able to build and retain talent. There is currently a significant lack of top talent in the market, and the CIO has a strong role in developing a positive and collaborative culture.	The CISO must also be able to build a strong and capable cyber team. There is a currently a gap of more than 4 million professionals in the field of cybersecurity, so there is a need for the CISO to be good at team building.
Ethics	The CIO requires the ability to carry out ethical decision-making. There are often trade-offs required and the CIO must ensure responsible and ethical use of technology and data.	The CISO's decisions similarly require ethics considerations, such as GDPR privacy concerns and sensitive data protection. There will be additional challenges around AI and LLMs to address.

Table 2.3 – Shared CIO and CISO soft skills

There are also specific soft skills that the CIO and CISO must separately address, and these are relevant to each role. These differences have more to do with their role than with the strengths of the individual. For a CIO, these include the following:

Business acumen	The CIO must speak the language of business and understand the business processes and how technology can create value, improve processes, and contribute to business growth.
Vendor management	There is a need for strong vendor management – the CIO must negotiate contracts, evaluate vendors, and ensure they deliver value as expected.
Innovation	The CIO must encourage innovation and an environment that helps the organization learn from new technology.
Conflict resolution	The CIO must have soft skills to resolve conflicts that invariably occur in the business of managing and driving change.
Change management	The CIO must have strong change management skills. Their role is to be a strong sponsor of transformations to adopt new technologies and business processes.

Table 2.4 – CIO additional soft skills

In contrast, some specific skills for a CISO include the following:

Collaboration	Cybersecurity is a *team sport* that involves various departments and difficult stakeholders. CISOs need to collaborate effectively with other IT teams, legal, compliance, and business groups, including their external customers, to implement security controls.
Negotiation	The CISO will regularly be required to negotiate with external parties, including vendors, partners, and customers. This is also required when there are serious incident responses that cross into other teams and departments.
Crisis management	The CISO is the captain of the ship when there is a crisis. When there is a cyber event, the CISO must remain calm under pressure and lead the incident response team. During this time, the CISO must effectively communicate with senior management and all external parties.

Continuous learning	The CISO must be curious and a lifelong learner. Cybersecurity is a rapidly evolving field and the CISO has to be a role model to their own team regarding professional development.
Stakeholder management	The CISO works with stakeholders across the business at all levels of the enterprise, including the board, executives, external partners, and customers. The CISO then must be able to manage these relationships effectively.

Table 2.5 – Additional CISO soft skills

Which of these soft skills are in your current personal brand? And which ones do you need to work on in your development? This is what we will explore next.

Understanding the gaps in your soft skills

Take the time to review the soft skills we've discussed and assess where you are with this. Which ones do you need to work on that are going to be harder and which will be easier to tackle? This cannot all be addressed immediately, so I would recommend that you identify one or two of these soft skills and add them to your SKEB analysis to tackle these gaps.

Here is an example of a SKEB analysis that I did for myself. I hope this helps you with using this approach to figure out what the areas to focus on are, and then which of these to concentrate on. In this example, I have identified four potential competencies, and would then select one or two of these to tackle first:

SKEB competency	Example
Skills	**Business acumen**: Be good at financial management to understand the nitty-gritties of the budget and be able to analyze which parts transform the business as compared to simply run the business.
Knowledge	**Adaptability**: Deepen my knowledge around vulnerability management best practices for critical infrastructure – in terms of managing the overall priority process and governance to provide board and management transparency on the risk buy-down of different approaches.

SKEB competency	Example
Experience	**Conflict resolution**: Have a positive experience in terms of being able to anticipate the next major conflict that will occur and address it with the respective stakeholders. To do this, I will have to complete a stakeholder analysis and spend 1:1 time with these parties. My objective is not to stop conflict from occurring, but when it does, to be able to manage it as gracefully as possible.
Behavior	**Strategic thinking**: Demonstrate a strong reputation for being able to lead the enterprise across new and emerging technology risks. An example would be to develop a white paper on different Generative AI LLM approaches and how this can be adopted securely globally and across businesses, considering current and anticipated regulations. Share this white paper with management and executives to influence and accelerate its adoption.

Table 2.6 – My soft skills gap analysis

Based on this updated analysis around soft skills gaps, you would circle back to the earlier *Table 2.2* and refine your SKEB analysis to add perhaps a few of these for further attention. The SKEB analysis is a living document. I would recommend that you update it at least one or two times a year.

Summary

The journey to becoming a successful CIO or CISO is a deeply personal one. At times, it is one that requires us to be uncomfortable and willing to learn. The volume of growth and the time required to complete it will depend on your own traits and the existing gaps.

There is no such thing as a perfect CIO or CISO. Having ambition and drive is an essential ingredient, but without a clear plan to get there, this will be a longer path.

This chapter introduced traits that the CIO and CISO commonly possess, and a few more that are more role-specific. If you can work on these identified areas and understand which soft skills require focus, this can accelerate your journey. In other words, the road to becoming a CIO or CISO will require you to self-examine your strengths and weaknesses. I have introduced how I use the SKEB model for this purpose.

I have worked with hundreds of CIOs and CISOs – they are all different, but many also share the same traits that have made them successful. These may not be the same traits that you have yourself, and that is absolutely fine. It is important to play to your strengths and develop ways to work on your weaknesses.

In the next chapter, we explore how to move from developing your plan to starting to execute one on your journey to becoming a CIO or CISO.

3

Executing Your Career Path to Becoming a CIO or CISO

This chapter aims to help you understand how to build a **career plan** to get you to your destination.

> **The fundamental question that is to be addressed in this chapter as follows:**
> How do I develop my career path to become a CIO and CISO?

There are conscious choices you must make regarding your career objective and position goal as you contemplate your career. A **career objective** and **position goal** are elements that I have included at the start of my own resume in the past. Doing so can help both you and your prospective employers to understand your career ambitions, and more importantly, how the role you seek is aligned to your long-term plan.

In this chapter, we will cover the following main topics:

- Developing your objectives
- Building a plan to make you grow and be *uncomfortable*
- Paths to becoming a CIO or CISO
- Exploring career approaches to progress
- Reviewing the CIO and CISO interview process

Developing your objectives

Having a clear and articulated set of objectives can be used for the right discussions when you are applying for a new role. Let's look at an example of career and position goals:

- **Career objective**: "To be a successful CIO at a dynamic and growing company where IT is a strategic component of the business strategy and my impact as a leader can be significant."

- **Position objective**: "To be able to make a valuable contribution as a senior leader in an organization that uses IT strategically and will allow me to develop and grow my leadership experience aligned to my longer-term career objectives."

You may be asking yourself how to take this example and apply it to your own situation. How do you feel about declaring your stance and ambition?

For example, to write that your career objective is to be a CIO or CISO might feel perhaps too ambitious to declare at the early stages of your career. Instead, you may write a more generic senior management role as the objective.

There must be some judgment call you have to make when you share this with internal or external parties that this document might be misinterpreted. For instance, your direct supervisor may be less ambitious than you are, and then you must judge whether your disclosure would mean that there will be more support for your career, or whether it could create unwanted tension as they think you want their role.

I would suggest including this information in your CV, but only sharing it and using it publicly in certain circumstances. The direct value to yourself is that it forces your alignment to this objective and makes you consider how any role that you might seek is aligned with it. There are certainly times when I would not declare this in my CV, and some personal discretion is required when you believe this may be inappropriate. In my own case, I used this format at the top of my CV as I had personal ambitions and wanted to execute my career plan.

The objectives themselves can be as specific and industry-focused as you consider applicable. For me, I prefer to not confine or ring-fence myself to banking or insurance. Instead, my preference is to note the key characteristics that I care about, such as growing, dynamic industries and contributing where it really matters.

Building a plan to make you grow and be *uncomfortable*

When I considered the CIO or CISO role as my personal aspiration, I would imagine myself in this position. Not in terms of the status or salary package that this would bring, but more in terms of how I would operate.

In *Chapter 2*, I asked you to explore areas where you could make improvements. There should have been at least two to three items on that list that make you *uncomfortable*. This is all about stretching yourself to get you out of your comfort zone.

If I reflect on how I used this myself, I can offer two examples:

- **Strategic thinking**: My natural strength is strategic thinking, but there was a three-to-four-year period where I had a specific focus on improving it. What I did was force myself to read 12 books every year on topics that were new or of interest to me. Once I committed to this objective, I made a simple reading log to record the titles of the books.

  I also recorded this personal objective within my annual objectives for each year. The broad learning from my reading was naturally incorporated into how I thought each day. Using regular reflection is a very healthy habit to establish and helps you to put things into perspective.

- **Attention to detail**: The second stretch goal was paying attention to detail. As I operate instinctively, I would find that I don't need all the data to make decisions. On the one hand, this meant that I didn't sweat about the small details and drive my teams crazy with requests for further information. However, I did learn that each organization and supervisor has different expectations.

  In a few roles, I learned that not knowing the fine details would be seen as a weakness. This was not too different from my home life, where my wife took care of this aspect – she is well qualified in this. At this point, I started applying extra focus to the details, particularly in the realm of financial management. On the work front, I had to learn that this would not work and my challenge was then to understand the fine details including the run rate, historical comparisons, and trends in my CIO budget.

Let's now discuss an example where working on these objectives helped me. When you are a CIO, there are times that you may inherit a large operating expense budget and it is not easy to understand the breakdown of these components. In fact, is this even something that you would want to focus on in the first 90 days of your new job? I cover this in *Chapter 5*, so read ahead or stay tuned. Having an MBA and an education in financial management does not mean that you are an expert. It just means that you understand some of the tools and lingo. What mattered in this case was attentiveness to the budget details and being able to manage the fine details when required.

The acid test came years later when I was in a CIO role and delivering on an $80+m transformation. The CFO in his wisdom had given me a significant target to take the existing **business-as-usual** (**BAU**) budget and allocate this to the transformation. Regardless of whether I felt this target was justified or not, I was able to meet this challenge without creating any new risks to projects or staff resources.

Another example I recall is in another CIO role, where the IT Strategy and the required architecture evolution roadmap had shown that there was $150m of obsolete technology. This constituted a high level of risk for cyber disruption and the availability of systems. The discussion and presentation with the CEO were unpleasant as, although I was new in the role, it was clear that four to five years of backlog was now being requested to be updated. In this example, I had to do a line-by-line explanation of why this could not be deferred any further. The following month, this $150m remediation budget was approved. While this was an uncomfortable exercise, it was what was required. When you are a CIO or CISO, it is often not a popularity contest; the requirement is to do what is needed and not what others may desire.

The key lesson I learned is to be curious and a lifelong learner. There are few individuals so talented that they don't have to continually *sharpen their saw*, and in IT this is particularly the case. The degree of digital and cyber change means you must always be in high-gear mode and go fast. Going fast does not mean that you should be *mechanical* though; instead, take the time to keep learning and looking for gaps in your knowledge.

This is another example of being *comfortable being uncomfortable* with the degree of digital and cyber change and being able to learn to challenge and accept which risks can and can't be accepted.

Now let's talk about your path to become a CIO or CISO.

Paths to becoming a CIO or CISO

The first thing to discuss is that it is likely that along this journey, the label of the actual role that you aspire to may change. The titles of CIO and CISO emerged at different times and there have been other senior labels that have been commonly used since.

The best advice I ever received on this topic was to think two roles ahead. Indeed you may have considered updating your CV with a career objective and position objective. There is usually a good chance that you are more than one job away from taking on the role of a CIO or CISO.

Let's look at an example of how you can plot your journey to this role and the various roles that you may have to take up on the way:

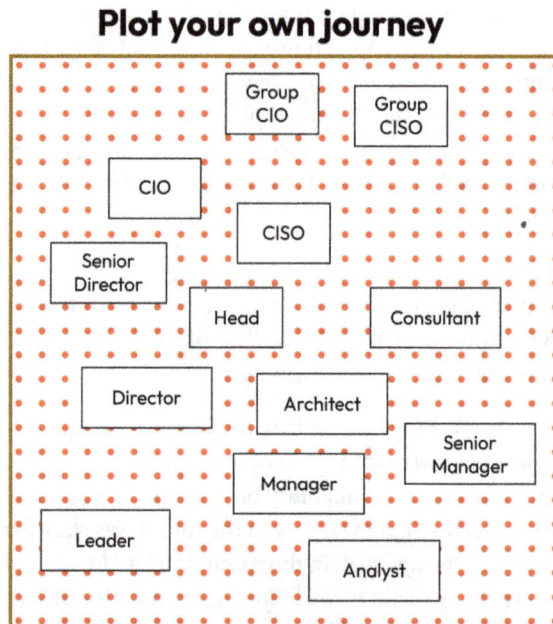

Figure 3.1 – How do you plot your path forward

There is no one path along which to progress to be a CIO or CISO. You should prepare a plan that moves you forward across different roles, but it is much more critical that you develop your leadership experience and behavior on this journey. This practice time will allow you to grow into more and more areas that were previously uncomfortable for you. Yes, you must have strong foundations in the requisite skills and knowledge to be able to make it to the CIO and CISO positions, but it is the experience and behavior that will accelerate this path forward.

As you think about joining the dots, then a CIO or CISO role might not be the end in itself – there are larger CIO and CISO positions that you can aspire to in different companies, industries, and locations. Then, there are further dots that the CIO and CISO should consider, as covered in *Chapter 11*, which may not be in your sights. But, I will introduce that as a consideration as well.

Your career plan does not have to be that complicated, so don't be afraid. There are patterns that will, however, accelerate this path so that this is not required to be a traditional step-wise path.

Earlier in this chapter, we mentioned that you should think two jobs ahead. What exactly does this mean? Let's discuss that next.

Thinking two jobs ahead

In your current role, there may be elements of experience and behavior that you won't get an opportunity to demonstrate. As you think two jobs ahead, you realize that this would be an expectation for those jobs. A classic catch-22 dilemma.

For example, leading an overall enterprise strategy refresh is not within the expectations for your current position, but you recognize that this is a gap that you need to address to be a fully-fledged enterprise architect.

There are **skills, knowledge, experience, and behavior** (SKEB) required to deliver an enterprise strategy refresh. So, how can you find ways to get this experience? Some ways forward include the following:

- Volunteering to be on the Working Group for Strategy Refresh
- Being engaged in a solution architecture review for your project and declaring your interest to a few peers
- Taking on a divisional or country-level architecture review
- Finally, being bold enough to ask for one-to-one time with the head of enterprise architecture to see if a secondment or involvement in the project is possible

A great example I have from my own career is at times moving sideways as an approach to moving upwards. I had summited a career path at one stage in my career and was CIO for a large pharmaceutical company – Eli Lilly. Many of you will perhaps be aware of this company; as of 2023, it had a market value of $495 billion, making it the largest pharmaceutical company in the US. Over a period of 15 years, I served as CIO for Australia, China, Asia Pacific, Japan, and the USA. These roles involved being an expatriate leader in each country, and the challenges were different in each one. On my return to Australia after years of being away, I realized that I had outgrown the healthcare industry and needed to find a new path.

A few other examples of thinking two jobs ahead come to mind: one time, I was a senior consultant at EY and thinking about a promotion to director and then either partner or CIO. I wanted to gain a reputation for developing robust IT strategies, which would be great SKEB for the CIO role. Another example is when I moved from KPMG back into a CIO role in financial services. This allowed me to re-enter an industry that I had been in back in my very first position. The new CIO role was also very attractive as it was a *transformation CIO* assignment to deliver significant change in a relatively fixed period of time and without the use of consultants. Delivering on a new core banking system and a payments, mobile, and online banking transformation while also changing all aspects of technology, cyber, and data was an offer too good to refuse. It meant that I would have to deliver on this agenda and learn on the go!

Both the CIO and CISO are apex roles with scopes covering the control of staff and budgets, and broad delivery responsibility. The staff are based in remote international locations and there are senior and difficult stakeholders to manage. There is also a strategic plan and the allocation of resources to address the business strategy. Vendor management and external industry collaboration are difficult, and there are regulators that require close attention. In this simple description alone, there are likely several areas in which you have not yet had the opportunity to gain experience or demonstrate leadership behaviors.

Thus, as you reflect on your current role and the opportunities in the market, both internal and external, the trick is to understand how this will help address your own gaps. Does this take you another step closer to the CIO and CISO role that you aspire for?

There is a caveat here: jumping from job to job is not what I mean. The key point is to be mindful that any potential shift is aligned with your career and position objectives and will address areas of growth that you desire. I've seen colleagues who change roles just for an increase in salary, but in the end, let's remember that you get what you seek. Take the higher road, and if that translates into the aspiration to land a CIO or CISO role, then seek the more difficult road.

Note that it is not required to be a perfect candidate with every base covered. If you focus on your *superpowers* and these are strong, then they will compensate for other gaps. The best way to get to CIO and CISO positions is to take on opportunities that allow you to demonstrate these attributes. There is incredible power in positive role-playing, as demonstrated in the following quote:

> *What you think, you become.*
>
> *What you feel, you attract.*
>
> *What you imagine, you create.*
>
> *- Lord Buddha*

Is there a secret sauce to accelerating your career to CIO and CISO? I believe this is sitting right in front of you, in the literal sense. The algorithm is not obvious, so let me introduce it to you next.

Introducing an algorithm to accelerate your own growth

In this section, I will explain why your own personal growth as a CIO and CISO is directly impacted by the strength of the team that works for you. When this team is strong, they can have a more significant impact, which positively improves your own reputation. The reverse is also very true.

As a leader, you inherit a team that has varied levels of talent. In my career, I have seen many examples of other leaders taking their old colleagues from job to job. I'm not criticizing this practice; this is quite common, and leaders want to trust and be able to delegate to a recognized associate.

I have noted that what I have seen is that your own growth is impacted by the strength of the team that reports to you directly. Your own ability to deliver superior results is impacted by the quality of these resources. This has tended to reinforce the pattern for leaders to bring their teams with them. However, the unintended consequences are that this builds divisions within your team. Instead, what I would propose is that you grow yourself by growing others. How can you do this?

Let's look at a simple construct to describe your team:

- Some members are 8 out of 10 in terms of their ability and contribution.

- Another may only be able to attain 6 out of 10.

- Each member interacts with others in the team, and in some cases, this is collaborative so that their efforts have a positive (+) or even a multiplicative (x) impact.

- However, some staff have poor teamwork skills, and their efforts can reduce your team's overall outcome. These become minus (-) equations.

This can be represented as follows:

$$8 + 3 + 9 \times 6 - 8 = ?$$

As the leader of this team, what do you do?

First, your role is to get everyone to perform at their highest-ever level. I often say, *be or bring the best version of yourself*. Thus, for this example, we want to get staff members at 8/10 to stretch themselves to be level-9 performers.

We also want to encourage the overall teamwork of staff members so that *trust* is maximized and *conflict* is minimized. I have applied the **Patrick Lencioni Model** to think through how teams overcome their current dysfunctions. Let's see what this looks like:

The five dysfunctions of a team

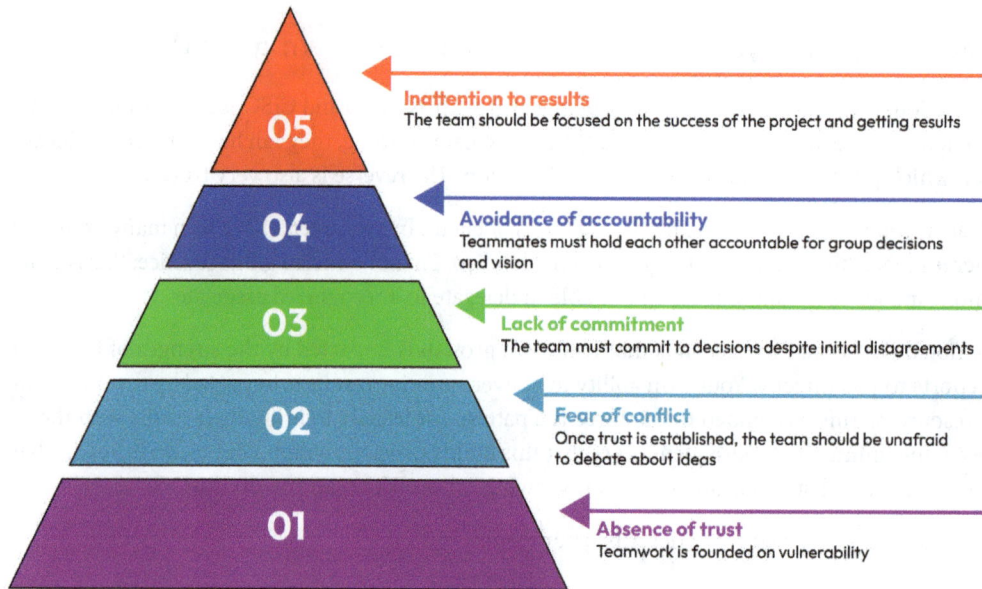

05 — **Inattention to results**
The team should be focused on the success of the project and getting results

04 — **Avoidance of accountability**
Teammates must hold each other accountable for group decisions and vision

03 — **Lack of commitment**
The team must commit to decisions despite initial disagreements

02 — **Fear of conflict**
Once trust is established, the team should be unafraid to debate about ideas

01 — **Absence of trust**
Teamwork is founded on vulnerability

Figure 3.2 – The five dysfunctions of teams

For more details, you can refer to Patrick Lencioni's *The Five Dysfunctions of a Team: A Leadership Fable*. By coaching each staff member through their SKEB with a strong focus on soft skills, you improve your algorithm. In summary, it does not require a team of stars, but instead that each staff member operates at their best and works effectively with others.

As the leader of this team, your own development will then accelerate as the team brand visibly improves, reinforcing your own personal brand. In our example, the starting point of the team was represented by the following equation:

$$8 + 3 + 9 \times 6 - 8 =?$$

After providing feedback, this equation was modified to the following:

$$9 \times 4 \times 9 \times 7 \times 8 =?$$

In this simple example, most staff responded to feedback and coaching and their individual performance has improved. You will also note that the interactions between team members have improved, and friction has been removed. This then allows stronger team synergy and output, hence better outcomes.

Often, I've seen leaders not do this, and I was always puzzled as to why. Was it that they did not understand the logic of this approach? Or were they themselves afraid that this development of their staff would actually put pressure on their own role in the future? To me, this is the natural role that a leader should play, and to get their team to operate at peak performance will require not only their personal best, but also that the different parts of the team work together at optimum levels.

In a relay race, the team with the fastest runners usually wins the race. The baton changeovers need to be practiced, and most athletes will focus on their individual 100m practice. In the 2016 Olympic Games, Usain Bolt won the individual gold medal event with a time of 9.81 seconds. There were no Japanese competitors in the final, and Ryota Yamagata ran 10.05 seconds to finish fifth in the semi-final. Despite these deficiencies, Japan won the Silver Medal in 4 x 100m men's relay, behind Jamaica, the favorites for the event. Both teamwork *and* individual performance matter. For relay races, the baton interchange requires the two runners to synchronize their change at their maximum speed.

Figure 3.3 – Great baton changes combined with speed gets you ahead

The next section will explore the different approaches to managing your career. These approaches provide you with some different routes to take up the mountain.

Exploring career approaches to progress

The philosopher Rumi said, "*What you seek is seeking you.*" So, when it comes to approaches to career progress, it is all about what you want to do and how willing you are to sacrifice to attain your goal. There are different career paths that you can take to the role of the CIO and CISO. It is not always the case that the same path will be trodden by every person. We are all different, and we need to appreciate the alternate routes you can take.

In this section, we will discuss the four different routes that you can take on your journey. We will note the *traditional* career path as a classic approach and the *linear* career path as an accelerated one with commensurate growth. The transitory and spiral approaches are less observed but are interesting alternative options.

Let's look at these four routes in greater detail:

- There is a **traditional career path** that we all understand. Your path will depend on moving through the ranks within your company and taking opportunities for development and growth. The traditional path for a CIO is illustrated as follows:

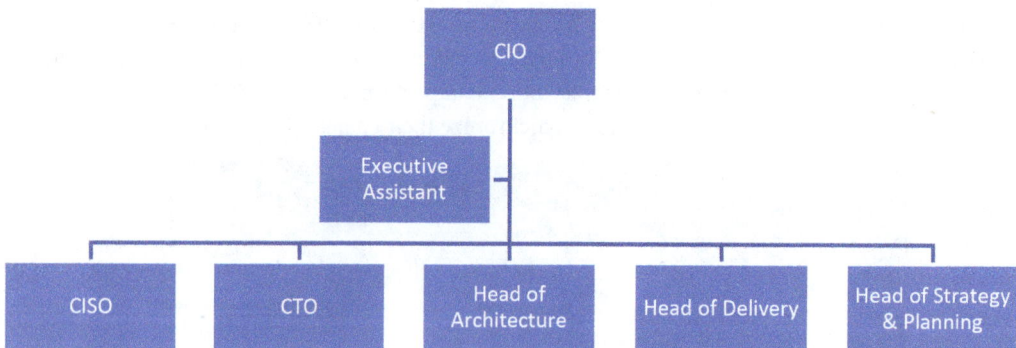

Figure 3.4 – Traditional and linear career paths

In the traditional path, you move from role to role, taking opportunities that look and sound attractive, but keeping a strong focus on continuity.

- The **linear career path** is more focused on rapid progress to move up. In this case, bigger titles, larger budgets, and more staff are what you seek. This also means that you start to collect a list of impressive titles. In taking this path, you ignore the organization chart and seek the fastest elevator upwards. You aim to move rapidly to new CIO and CISO roles, so in that sense, it is the same as the traditional path, but with greater velocity.

- A **transitory career path** is one where you make it a rule to continually change from role to role frequently. This makes you very adaptable to change, however, you may not be able to demonstrate a strong track record of delivery. This path will often traverse different companies, industries, and perhaps geographical locations.

Transitory

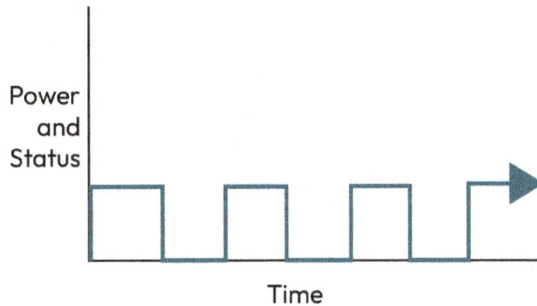

Figure 3.5 – Transitory career path

- A **spiral career path** means making lateral changes frequently to gather new skills and experiences. This may require you to move between different departments within IT or between business and tech positions. This may also mean that you move backward to move forward. In this path, not only do you ignore organizational boundaries, but you also move more broadly than what is normally expected in a career.

Spiral

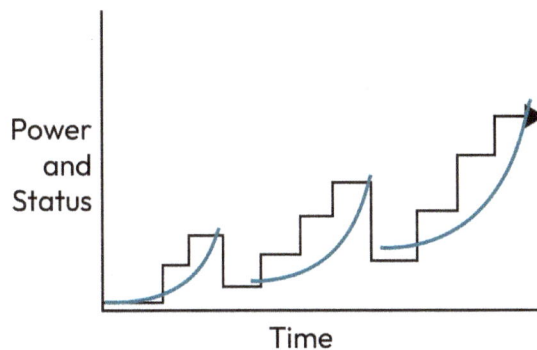

Figure 3.6 – Spiral career path

The following table provides a summary of these career development approaches. We are used to naturally thinking about career development in the traditional or perhaps linear senses. Your own path can even include a few different paths if you sense that this will give you an advantage for growth.

Career path approach	What this looks like	How this feels
Traditional		A slower and more steady approach to building your career. This is purposeful and without many surprises. You are building a bank of trust with stakeholders.
Linear		Fasten your seatbelt as acceleration is the goal. It may get bumpy at times. There will be times when your role exceeds your readiness, so *learning how to learn* is critical. Also, be sure not to make enemies on the way up.
Transitory		This should feel comfortable with some stretch as you look to move around laterally to learn and gain more experience, with the expectation that this will provide more upward mobility at some point in time.
Spiral		You keep learning as you move roles, and this may mean going backward to go forward. It can also mean lateral rotations. Either way, you need to have confidence in yourself.

Table 3.1 – Career routes – not always upwards

Your career path will be a combination of these routes, and at different stages of your career, you may seek or desire a career to match your overall aspirations. Taking riskier paths such as the linear or spiral paths can be harder to accept in the middle stage of your career, for example, if you are supporting a family. Moving to a transitory path from a traditional steady path will be easier to adapt to and perhaps more suitable.

Again, remember that we are all wired differently in our ambition for the CIO and CISO roles. Similarly, once you have been on a linear career track for a while, it is also hard to get out of that mode and adopt a spiral path or any other approach.

Our discussion on the path to becoming a CIO and CISO would not be complete without discussing the interview process for this role. Let's look at this next.

Reviewing the CIO and CISO interview process

You have your CV prepared, so what other preparation should you complete? The thought of being interviewed should be an event that you look forward to, rather than one of doubts and anxiousness. This section explores how to find success through the interview process to get your promotion to CIO or CISO. The interview process to become a CIO and CISO has many variations and will depend on whether it is an internal or external hire.

The external CIO/CISO role

Let's say you have identified an external CIO or CISO role that has been advertised, or even better, you have been head-hunted by an executive search firm. For significant CIO and CISO roles, while some may be advertised on LinkedIn and other job search sites, they are typically placed via recruitment firms. The recruitment firm will usually have exclusive rights to this position and will be paid a retainer to locate the most suitable candidate. Let's first look at how these recruitment firms work.

How recruitment firms operate

Receiving a phone call from an unknown number can be both daunting and interesting. Recruitment firms are well-networked and have access to information about potential candidates in adjacent companies. The old saying is very true – "*having a job is the best qualification to obtain another role.*" They also have international colleagues who provide input to CIOs and CISOs that are perhaps open to a move. Often, recruitment firms also receive recommendations from other colleagues in the industry that are normally anonymous. Again, this leads to more intrigue.

A common response is "I'm currently enjoying my role and not presently looking for a job". Regardless of how you choose to respond, always be polite. In such a scenario, I would respond by clarifying that my career objective is to be CIO/CISO and while the offer is intriguing, I would still want to understand a bit more about this opportunity.

One key thing is that you want your personal brand to be well respected by these recruitment firms, so take time to develop these relationships over coffee. Even if this opportunity does not work out, then they know what SKEB you bring.

Recruitment firms are financially motivated to place you in the role as they generally obtain a percentage (25–35%) of the successful candidate's salary in their first year. These fees are usually paid in three equal installments. In some cases, recruitment firms also conduct psychometric assessment tests on candidates.

Recruitment firms will provide you with the job description and an overview of the challenges of the CIO and CISO role. They are unlikely to inform you of the company's name, and this will be hidden until later. In many cases, this is confidential as the current CIO or CISO is still in the role but the company's management has decided to seek someone else instead.

The screening process can be quite lengthy and require many months as the potential candidates are in jobs already, so they are not always willing to take a random call. My advice is to be patient and not get too anxious; it is hard if you really want a job move, but the process can only go as fast as the hiring manager pushes this.

Once you have agreed to be a candidate for the role and the recruitment firm has formed its longlist and shortlist, then the recruitment firm will be able to inform you of further details about the company, the incumbent CIO/CISO, and what their client is searching for in terms of the replacement.

Doing your own research

For an external role, you must complete your own research on the company and understand as much as possible about the CIO position in that organization. The details of listed companies can be easier to get hold of as they publish annual reports and analyses of their businesses. My suggestion is to prepare some written notes, as this will force you to read and interpret widely available information. This may include the following questions:

- What are the vision and mission of the company? (Does this align with your own views?)
- Which sector of the industry does it operate in? (Is this a growing or mature industry?)
- What products and services do they create? (Is this aligned with your personal values?)
- Where does the company operate? (Consider the size and complexity of the role.)
- What are the priorities (i.e., strategic drivers) for the CIO/CISO?
- Do you understand what issues and challenges will be faced?
- How long was the tenure of the current CIO/CISO (noting that the average tenure of a CIO is 4.5 years and a CISO is 2 years)?
- What is the interview process? (Consider the number of interview rounds, panels, socialization interviews, etc.)
- Who is interviewing you? (Consider the backgrounds of the interviewers and stakeholders.)

Next, let's discuss the value of role-playing before an interview.

Interview roleplay

Unless you have recently had an interview, then consider taking time with a trusted friend or partner to run through the questions and your own background. Take the time to practice and ensure that your own brand comes through. In the interview, you want to be confident but not arrogant. You want to be informative but succinct. The opportunity to do a *dry run* is very important. I recall having to do this in one CIO role I was interviewing for. In this case, I had flown to Tokyo for two days of interviews with the CEO and management team and had then been informed that, on the last day, there would be an interview in the Japanese language.

While I never advertised myself as being fluent in Japanese, they wanted to test my level of conversational-Japanese-language skills. To tackle this properly, I then did a dry run in Japanese with some friends (Greg and Kathryn) who lived in Hokkaido. This meant that the next day, I walked into the session with a positive outlook. I even practiced my yoga that morning and, in my mindfulness practice, had chanted, "*They want me to succeed.*" Positive thinking plays an important role in your mindset when you want to take on a CIO and CISO role.

Now let's discuss what you can expect in the interview itself.

What to expect in your interview

I've seen many variations of interviews – one-on-one, two-on-one, and a panel-of-10-on-one. For one senior CIO role, I had 17 interviews, including being flown for two days to an international site for the process, while the latest interview I did was with a Global CIO and that was at midnight.

The interview process can be both hard and tiring. Preparing yourself mentally to be at your best is critical: you want to ensure your energy levels and communication are as crisp as possible. Being lethargic and withdrawn won't be in your favor. It must be remembered that you are interviewing them as much as they are testing you. Therefore, you must prepare questions for each interview and as far as possible, tailoring them to what you understand to be the key areas of interest.

There is a detailed analysis of questions that you may be asked and what you can ask at the interview in *Chapter 4*. In your career as a CIO or CISO, it is to be expected that looking for the next elevator will be required every few years. This is very rarely a one-and-done situation.

Some companies use a structured interview process where some key dimensions of the CIO and CISO role will be explored, and you will be asked to "explain a time when you…". These will focus on the *superpowers* (soft skills) that I outlined in *Chapter 2*, including leadership, strategic thinking, problem solving, communications, adaptability, and emotional intelligence.

Now that we've looked at the external interview process, let's look at the scenario in which you're an internal candidate.

The internal CIO and CISO role

This process will be radically different, as you are a known entity. Many of the aspects of research will not be required. The first fact to establish as an internal candidate is whether external candidates are also being considered. For internal candidates for CIO or CISO roles, these people will be tapped and asked to apply. There will also be times when you are interested in the CIO or CISO role as it is vacant, however, you are uncertain of whether your chances are high or not.

Preparing for the interview

Taking on internal CIO and CISO roles can indeed be harder as you have an established brand. The CEO and COO interviewing you will need to be able to assess your potential to operate on that broader and more strategic level required for these roles. They are used to seeing you operate in narrower focus areas, so you will need to work harder to reset expectations. To this end, I would recommend the following:

- Conduct private research to understand the largest issues that the CIO or CISO were unable to address.

- Find out who were the major stakeholders that were not supportive.

- Assess whether you are seen as being too close to the prior incumbent.

- Understand what you bring to the table. While you may not have existing CIO or CISO competencies, is there some experience (e.g., transformation leadership) that puts you in a good position and can be leveraged in your new role?

- Have clarity about your own vision for IT and cybersecurity.

- Gauge how your current colleagues and peers will react to you being appointed.

Once you have evaluated that there is no sensitivity in terms of your application among the relevant stakeholders, I recommend you declare your interest in the role and seek their input. The way to frame this is that you value their insights and are aware that you had worked with the prior incumbent, then asking yourself the following:

- Do they have any input on the strategy and what is missing?

- What do they see as areas of improvement?

- Why do they think the prior incumbent was not successful?

- Is there any specific feedback that they can share on how to position myself as a strong candidate?

Just as with the external CIO and CISO candidate approach, prepare questions to ask in the interview (see *Chapter 4*). These will have to be somewhat harder and less superficial as you have inside knowledge of current gaps. The questions may therefore be used to test some hypothesis around areas that you believe could help the stakeholders and any issues that are being experienced.

As an internal candidate, you should also be ready to answer questions about your candidacy. Typically, this would cover key areas of the CIO and CISO responsibilities such as the following:

- Development and execution of the tech/cyber strategy

- Your strong business network and examples of how your understanding of the business can be an asset

- Any track record that you possess in driving transformation change

During the interview process, don't be tempted to pad and tell *white lies*. It is easier to just focus on what you bring to the role, and where you need to develop.

You are now very close to getting your dream job. But retain your confidentiality about the opportunity and try not to let your thoughts get distracted in your present role. It is natural to see things through a different lens and start to think how others will react once this is announced, but my advice is to not allow your mind to wander there yet.

Now let's assume you've been informed that you've been selected for the role. What should your next steps be?

Selection as the preferred candidate

Congrats! You have been informed that you were selected, perhaps followed by a fist pump, and then there will be some further discussions around the exact offer and terms of the contract. The negotiations should always be pragmatic and positive.

The offer is always somewhat different from what you expected, with different structuring and ranges of salary, bonus, shares, and so on. No two companies operate the same way in terms of their packages offered and usually there are only some factors that can be negotiated.

It is always best to talk over your offer with your spouse, partner, or closest confidant. My guidelines are that you want to be in the *ballpark* of a salary range, rather than be insistent on an exact number. The HR recruiter you are speaking to usually has some flexibility but must operate within the guidelines (salary scales) set by the compensation manager. Note that any exceptions to the new package will go back to your potential new supervisor, so be aware that being too needy or demanding may contribute things to your personal brand that you don't want.

As the preferred candidate, you have a verbal offer, and a written offer will only arrive once you have finalized all the terms and conditions. The written offer will be subject to reference checking and most likely also background checks.

For internal candidates for CIO and CISO roles, this is a joyous occasion and congratulations are certainly in order. The terms and conditions are usually less formal and there may be a transition period before an actual promotion is completed. The major difference is that if you get the role, then it is likely that some of your peers and colleagues will now work directly below you. This will be a degree of adjustment and will require some mature reflection to ensure that this reset is established.

Summary

It is a deeply personal journey that each of us must undertake to make this voyage to our career goals, one that requires being uncomfortable and stretching ourselves. The growth that will be required will depend on your own natural traits and what gaps exist. To execute your career plan to be a CIO and CISO, you must be a lifelong learner and remain hungry. You can connect the dots by thinking two jobs ahead; this will allow you to find the roles that will develop specific skills you need or provide you with experience that is on your development plan. Become an expert in the simple algorithm that allows you to grow yourself by developing your team. This is not about being the star runner; it is the combination of the members of the relay team that matters. Understand and commit to career approaches to progress at different stages of your growth. Each of the career approaches covered here may be used at different times of your career.

The interview process for CIO and CISO roles will probably be the hardest and most complex that you have ever experienced, particularly when you are an external candidate. As a result, the preparation required should match this level of rigor. For internal interview processes, this can be very different as you have an existing brand, and you will want the interviewers to see you through a different lens. This will mean that you also must prepare well for the interview.

The next chapter is a guide to interviewing for your next CIO or CISO role. The chapter is a guide to preparation for the interview – in particular, what questions you should expect to be asked and then what questions you should ask when it is your turn to probe.

4

CIO and CISO Interview Tips

The interview process is unavoidable and is a necessary stage to pass through for the dream CIO or CISO role that you covet. There can be frustrations with how slow this can be and it can take place in a protracted or seemingly uncoordinated manner. Keep this as a mental note for when you are the CIO or CISO, and consider how you might want to run the interview and selection process. I've found it to be useful to put myself in the shoes of the candidate, and this has become part of my brand. Being transparent, responsive, and honest in the interview process sends a clear message about how I normally operate. In my experience, candidates get a flavor of what it would feel like to work for me.

> The fundamental question that is to be addressed in this chapter is as follows:
>
> How can I nail the interview for a CIO or CISO role?

Now let's get back to your preparation for the interview(s). There will be many dialogues, and the number of actual interview rounds can vary dramatically from three to even eighteen. I'm sure that eighteen is not the world record. Yes, I've been personally involved in a selection process that required eighteen interviews!

Preparation is everything and it does not matter how strong your CV and background is. If this interview session doesn't go well, then you will never get the position.

This chapter is useful guide to assist you in securing your next CIO or CISO position, which can be used by both candidates and also management teams charged with carrying out interviews for these role. When you consider the strategic importance of the CIO and CISO roles, it should come as no surprise that the interview would be a comprehensive and challenging part of the process. However, it should not be adversarial.

Ideally, the interview provides a mechanism for you, the candidate, to be well prepared for the role and engage in a process to assess whether you are a good fit and that the company culture is aligned with your expectations.

You should not expect the interview to be a simple walk in the park. The interviewer wants to ensure that your skills, knowledge, experience, and behavior are a good match and that you will be able to take on the challenging nature of the CIO or CISO role in their company. The interviewer will be assessing whether they can see you as the incumbent in the actual position. Yes, that will perhaps come with some existing bias, but that's how the process works.

I have been in interviews where there is a clear *bad cop* and *good cop* approach. One interviewer is friendly and engaging, while the other is much sharper and testing. The questions from the *bad cop* are much tougher and more challenging, maybe to see how you respond to solving a problem. Noting how you might struggle with a difficult question is a good test of your own ability to remain calm and think strategically in a crisis.

When you are faced with this situation, it is best to maintain a balance in your answers, giving equal attention to both the *good cop* and *bad cop*. This is why preparation for the interview is the key to success.

In this chapter, we will be covering the following topics:

- Prework and orientation
- The interview
- CIO questions that you may be asked
- CIO interview questions for you to ask
- CISO questions that you may be asked
- CISO interview questions for you to ask

Prework and orientation

Significant prework will be required to have an understanding of the company or entity that you are considering, usually involving some research around their operations, locations, financial statements, the board, and the management team. Some insights may also come from the recruitment agency that is involved in the process.

I suggest that you review some relevant news stories from the last few years. It is often difficult to gain an understanding of the IT or cyber strategy of an organization from news sources as CIOs and CISOs can often be quite low profile. There are usually some interviews or articles by vendors about successful new innovative projects that can provide limited insight.

As a minimum, I would suggest that you read the following:

- **Job description**: There are clues in the **job description** (**JD**) as to the company and its culture. The JD also provides an overview of the requirements for the role and who the chosen candidate will report to. There may be some clarification questions that you want to ask. This may also be provided by the recruitment agency.

- **Annual report**: The annual report provides a summary of the strategic priorities and the company's business with performance across the reporting period. You may be able to glean some questions from this report to ask in the interview.

- **LinkedIn profiles of key management personnel**: The annual report provides the bios of the key management and board personnel. However, it is usually of interest to understand the human network and who in your list of contacts already has relationships with any of the stakeholders.

- **Company website**: By exploring the company's website, you can gain some general knowledge of the company's history, future direction, and its products. Again, this is also a good source from which to prepare your own questions.

- **Company review sites (Glassdoor, Indeed, Workable, etc.)**: These sites provide an independent view of the company's reputation from other candidates and recent interviewees, offering a guide to the company's culture, compensation, and even the interview process.

Start your preparation early. This research should be done a few weeks prior to your interview. Where possible, check your LinkedIn to see if you know someone in the prospective company. An insider is usually a good source of information on the interview process and might be able to give you some tips.

Be sure to take notes to ensure you recall what questions were asked and what was discussed. It may be confusing when you get to the end and try to recall each conversation.

You should prepare for their questions and also prepare your own questions to ask the interviewers. These are both covered later in this chapter.

The interview

CIO and CISO interviews are often panel interviews with two to five interviewers. For these interviews, you will normally be seated in the middle directly across from the interviewers. My advice is to try to engage each of the interviewers equally with discussion, questions, and eye contact.

It is natural to want to talk and engage with your potential new supervisor. The other interviewers are likely to be key stakeholders or senior staff from the team. Be sure to listen and understand what their needs are and ask what they believe should be changed.

Some advice: there can often be an interviewer who has a poker face and does not smile. Their mannerisms may appear to be less friendly, and they are not as engaged. It is easy to make the assumption that they are not supportive but be careful as you really don't want to second guess their intentions.

Be sure to treat each person who interviews you with respect. I've also had multiple occasions where I have been interviewed and a member of my future staff was on the interviewing panel. Again, recognize and adjust yourself as this can be normal for some companies that want to ensure cultural fit. At the end of the interview, you want future staff members to feel that they want to work for you and that they can learn from this experience with you.

What we will mention in *Chapter 9* on the subject of stakeholder analysis applies here (you, of course, have not yet completed this analysis). Be respectful. Ask challenging questions, not too challenging though, but thought-provoking. I can't imagine that someone who is unwilling to engage will ever be hired as a CIO or CISO.

Your branding (discussed in *Chapter 1*) is very much on show in the interview process, and this is what the interviewers will take away. Their summary of the interview will usually be short and cover the following:

- An overall impression of you
- Any strengths and weaknesses
- How you addressed the questions
- An analysis of whether you're fit for the role
- An analysis of whether you're fit for the company culture
- Concerns they may have about you
- Recommendation on the next steps

The next step may be to halt the process or continue with a recommendation that you will also meet other interviewers. There may also be one or two items noted for further follow-up and questioning.

Following this step, the Interview feedback is communicated to HR and sometimes the recruitment agency. It is often not that descriptive, and the key message will be whether they want to continue the process or not.

I've learned to not ask too much about the overall process and how many interviews will be required, as companies often like to adjust this depending on the overall progress of the process with other candidates and what degree of socialization interviews are required.

As I stated initially in this chapter, the interview process can be frustrating and also give you a negative impression of the prospective company. It is best not to let those thoughts get into your head and impact your conversation. Now, let's explore what questions you could be asked in the interview.

CIO questions that you may be asked

These are a number of the standard and more complex questions that you could be asked. I've penned a few bullets to outline some example answers. I haven't written out any long-form answers, as when you are asked these questions you should never simply read the response. What is more important is to answer confidently and fluently and try to cover as much as possible. Remember that eye contact and engaging your interviewees equally are necessary. There is no point in having one interviewer vote yes for you and another vote no.

The following 25 questions I have either fielded myself or have heard asked. In preparation for your actual interview, I would suggest that you write some of your own notes for each of these example questions. These draft responses are only meant to be a guide to how I would answer this. It is always better that you do this in your own voice, and that you be totally honest in your responses. As I said, body language is critical – it is not so much what you are saying verbally as it is how you project a positive image to the interviewer.

1.	**What will be your strategic priorities for the next year?**

Note that you have a few key points here:

- Refer to your 90-day plan where you have a number of areas to focus on initially, including the IT strategy and assessing the team

- The strategic priorities will be aligned by building an understanding of business plans

- The stakeholder engagement will also provide some inputs

- Reference your support for any known strategic or regulatory requirements

The key is that you want to be able to articulate an IT strategy and demonstrate that you are comfortable with the complexity of this domain – that while you have a strong strategic mindset, you are open to listening and learning. The CIO who comes across as arrogant will be avoided.

2.	**What culture will you be trying to build? What do you see as the biggest challenge to doing so given the current environment?**

Acknowledge that the culture in the company is one of the major attractions for you in wanting to join. This should be aligned with your own personal brand analysis. But here are some tips. While answering questions about building a culture, you must touch on the following points:

- A diverse team that is willing to challenge and drive change

- An empowered team that acts with accountability

- A team that is aligned with business and IT objectives

- A data-driven team

- A learning culture built on curiosity

The key is that you have a passion for developing a culture that is going to add to and enhance the DNA of the company.

| 3. | **How will you engage with your key stakeholders?** |

Refer to your 90-day plan, which is part of your plan. Talk about the following:

- What are their key issues?
- Where would they like support?
- What would you like IT to do differently?
- Listen to their responses
- Take action on any quick wins
- Engage them in the IT Strategy Review

The key is that you have developed a 90-day plan that you want to guide your actions, but that you are still open for input and refinement.

| 4. | **How will you balance digital innovation and technology resilience?** |

Acknowledge that this will be a primary responsibility of yours. Remember to do the following:

- Acknowledge the importance of both and the need to make trade-offs
- Reference any specific enterprise change that is prompting this question
- Note that there will be times when counsel will be required from the CEO/COO etc.
- Note that the revised IT strategy should attempt to address this roadmap plan
- Talk about an example

The key is that you as the CIO have a measured approach to try to gain a good balance between these two objectives.

| 5. | **What is the biggest tech challenge to manage? (Generative AI, Cloud, DevSecOps, etc.)** |

Note that disruptive tech challenges will continue. These are some things you can talk about:

- Navigating change requires lifelong learning and critical thinking
- Taking an outside-in view (learning from best-practice sectors and companies)
- Horizon scanning by the CIO and their team will be ongoing
- Again, give an example where you did this successfully

The key is that you have strong subject matter expertise in the latest tech that is being used or planned to be deployed and that you demonstrate your commitment to sharpening your own sword.

6.	**How will you handle conflicts in key programs and with the team?**
	Acknowledge that for transformation, conflict is a natural outcome. But you can talk about the following:
	• Expect teams to be respectful
	• Understand the issues and facts
	• Try to find resolutions that make sense to all
	• Be transparent to share more broadly how conflicts are resolved
	The key is that you come across as a pragmatic non-political leader who wants to drive business outcomes.
7.	**Share with us your proudest achievement at your previous organization.**
	This is a free-hit opportunity to talk about a complex situation that you led to resolve. Address the following:
	• Describe the scenario and why this was critical
	• Describe what you did yourself to lead this
	• Note how complex and difficult this was
	• Summarize the outcome and any ongoing benefits since then for the IT team or business
	The key is that you are able to provide a scenario that you can talk through that is complex but can be explained in simple terms and demonstrate your leadership.
8.	**What are your reasons for wanting to leave your current role?**
	To answer this question, the focus should be on the new opportunity and not on any criticisms of the current one. Talk about the following:
	• Excitement and challenge of this new CIO role
	• The culture of the new enterprise is attractive due to ABC
	• The timing works for you
	• An opportunity to grow further
	The key is to talk about this new opportunity. It is exciting due to the challenge and what you feel that you can contribute to the role.

9.	**What kind of a leader are you?**

This gives you a chance to talk about your personal branding, which we have talked about in Chapter 2. For instance, a CIO's soft skills could include the following:

- Problem solver
- Strategic thinker
- Able communicator
- Talent developer

The key is to talk about your strengths and how they relate to this role.

10.	**How do you stay current with technology developments?**

You must answer this question in a way that shows that you are learning every day and that lifelong learning matters to you. You could say that you enjoy the pace of change and stay current by doing the following:

- Being curious
- Networking within the industry and beyond
- Reading broadly
- Taking time to research some specific tech, cyber, and data hot topics

The key is to talk about your commitment to life-long learning and how this keeps you current with some examples provided.

11.	**What do you see as your biggest challenge to being successful in this role?**

This is an interesting question – as you should be confident but also humble. My thoughts are to cover strategy, partnering, and the team:

- There will be strategic trade-offs that will create tension
- Build strong relationships with stakeholders in this environment
- Develop a team to support this change and to be *comfortable being uncomfortable*

The key is that you note that these are the known challenges, however, the nature of the CIO role is that new unknown ones will emerge.

12.	**How do you plan to develop your new team?**

State that your first 90-day plan includes this. Be sure to mention that this is also aligned with the refresh of the IT strategy so some of the details will be refined as you commence the new role:

- Assess the team (evaluate gaps)

- Build a future team (new capabilities)

- Get input from your boss and from stakeholders

The key is noting your structured approach to team development and your commitment to the algorithm to accelerate your own growth (*Chapter 3*).

13.	**How will you engage in the transformation program?**

Assert that as the new CIO, your role will be to help drive transformation and change. You could touch on the following points:

- Understand the transformation scope

- Build relationships with the key stakeholders

- Ensure that the refined IT strategy incorporates support for this transformation

- Have yourself and your leadership commit to driving this change

The key is they must see you as a leader and also a team player. If the transformation has already commenced, you might want to convey at the onset that you are unsure of the actual program status. But you could list the issues that are to be taken care of. They will want to see that you are engaged and see the transformation as a priority.

14.	**When you take on this CIO role, what will worry you at night?**

Let me suggest this response. You often don't have all the facts to determine the current biggest issue – as the new CIO, I have to trust the team, but as I'm new in the role there will be much that I worry about:

- Can we develop our team fast enough to meet the new business demands for transformation?

- Is the IT strategy taking on too much?

- Is the cyber strategy and roadmap realistic and achievable?

- How do I balance the short-term issues with longer-term objectives?

The key is that they want to hire a CIO who can manage complexity, is strongly committed, and is also resilient enough to be able to handle stress.

15. What is the best way to measure the success of the IT team?

There are many ways to measure success. My suggestion is that you talk about some of your own personal metrics that you monitor and note that this will be reviewed in the first 90 days. Here are a few I would suggest:

- Impact delivering value

- Customer satisfaction

- Financial responsibility

- The team operates as a collaborative and cohesive unit

The key is that you have a broad perspective and look from a business technology point of view. There is no wrong answer but too much focus on tech will be considered narrow unless there are current tech incidents on stakeholders' minds.

16. What is your most proud achievement?

Be sure to have two or three examples ready for this answer. Not having proud achievements would send a funny message. Here are a few examples:

- Completing your first enterprise-wide strategic transformation

- Delivering on the core banking transformation (within budget)

- The pride of seeing talent that you have coached and mentored become CIOs themselves

The key is that you are able to articulate a track record of successful delivery and are able to talk about the business outcome and not just the technical aspects.

17. How do you work to ensure risks are managed in the enterprise?

My suggestion is that you acknowledge the importance of risk management and how you would focus on this in your CIO role; in particular, the following:

- Ensure risk management is data driven

- Ensure risks are governed appropriately

- Ensure actions are proportional and weighted based on risk

- Ensure clear ownership and assignment of actions

The key point is to demonstrate that you prioritize risk management and have a sensible approach to addressing the most critical risks.

18. **How are IT programs and projects prioritized?**
My suggestion is that you refer to your 90-day plan. There may be some other current concerns that have led to this question. It may be worth enquiring whether there is an existing issue. I would suggest that you note the following: • The IT programs and projects are aligned to this IT strategy • Management and the board (as required) are engaged to prioritize these • IT portfolio governance is put in place to manage this delivery • Benefit realization is used to measure how value is achieved The key is that the focus is clear management engagement and governance to prioritize all projects including IT programs.
19. **What will you do in the first 90 days in the job?**
Share your 90-day plan in the interview. Note that the plan may require some fine-tuning after being onboarded and understanding further details. Also reinforce that this provides a rough outline of what you will do, of course, subject to input and feedback from yourself and other stakeholders. The key is that the plan does not become the interview. You are fine to have this tabled and reviewed afterward as this demonstrates your proactive preparation.
20. **Explain to me a recent digital innovation program that you delivered successfully.**
Select an example that is not sensitive. Here are a few points: • Explain why this transformation was important (business growth, merger, etc.) • Explain your role as leader or co-leader • Detail how challenging this program was (complexity, time constraint, etc.) • Provide an example of something that was very difficult and was overcome • Finally, summarize any lessons learned The key is that you can talk about the business benefits and how this innovation had positive customer engagement.

21. Talk to me about a situation when you were faced with large budget cuts.

Take an example where there were significant budget trade-offs required to meet an overall business objective. You can talk about the following, but complement your answers with the rationale behind them:

- Reducing short-term contracting
- Reducing consultant spend
- Cutting discretionary expenses (travel, training)
- Stopping projects
- Negotiating to protect staff expense

The key is that you can demonstrate your attention to detail and also your ability to strategically align with a business imperative.

22. What is your weakness?

You can refer to your SKEB analysis, as talked about in Chapter 2, as a reference. However, I would suggest that you summarize this:

- Disclose an attribute that won't disqualify yourself from the CIO role.
- Share what are you doing to address this gap.
- Position a negative as a positive – I'm naturally strategic and look at things from a bird's-eye view. I have been working on being conscious to force myself to embrace details.

The key is that you are acknowledging areas you need to work on and how you are progressing this weakness into a lesser concern.

23. Tell me about a time that you had a crisis to manage, and what you did.

The CIO has to be calm and structured in a crisis. The following are some key points to highlight:

- Outline an example in a recent role
- Note how the preparation and dry runs were useful
- Explain the process and the risks involved
- Show how you were decisive and a clear communicator

The key is that you can demonstrate your natural leadership ability to take charge in a crisis and resolve complex issues.

24. **Talk about an example when you had to resolve a heated conflict.**
Explain the scenario and the parties involved. Here are a few key steps: • Talk about what the *circuit breaker* was to allow everyone to cool down • Highlight that the approach was to focus on the problem, not personal attacks • Agreement on common areas • Move on with the solution • Follow-ups to ensure that all parties were satisfied with the final outcome The key is to first note that this was an unusual situation, and the leadership role that you played to bring this to resolution.
25. **Why do you want this job?**
This question is quite obvious, but here goes: • The company culture is attractive • The strategic direction of the company and how IT can contribute • The exciting opportunity to build new value through XYZ program/project • Finally, I'd really like to work for you (if that is really true) The key is answering this question with strong eye contact and speaking about how you are aligned with several factors.

In this section, we have explored the questions that you may be asked when you are interviewing for a CIO role. These 25 questions of course are not exhaustive but typical ones that could be posed. The key is that you are ready for these and any additional ones that you think could be asked. The next section then transitions into the second part of the CIO interview when you can ask questions back.

CIO interview questions for you to ask

You have reached the stage in the CIO interview where the interviewer then asks you what questions would you like to ask of them. Having no questions is never going to be seen in a positive light. Conversely, asking too many questions may also not be ideal.

My advice is to ask three to five questions and have these prepared prior to the interview. It is acceptable to reference your notes as this can demonstrate that you are taking this role very seriously and have come prepared.

I'd also add that this stage of the interview is where you can probe and gain a deeper understanding of the CIO role. Don't be afraid to ask a secondary follow-up question and be curious without being overly annoying.

Here are 20 questions for you to consider:

1.	What are the current key success metrics by which the IT team is measured?
2.	How well is the IT strategy understood and accepted across the business?
3.	What is the process for budget trade-offs across the enterprise?
4.	Are there any expectations of the board for the new CIO?
5.	In your opinion, how should the current culture be improved?
6.	Is there strong alignment across the business on IT strategy?
7.	What is your assessment of the IT team's capability and are there any known gaps?
8.	How is our relationship with key regulators?
9.	Are there strong advocates for digital innovation across the business?
10.	Is the cyber risk culture good enough? Or is further work required?
11.	Are strategic decisions all top down?
12.	What are the longer-term business expansion plans that have IT dependency?
13.	How can employee engagement be improved?
14.	What is the staff retention rate? What is the regretted leavers rate?
15.	What are your expectations of me in the first year?
16.	What is the largest challenge that we face for business growth/ transformation?
17.	What surprised you about this company when you first joined?
18.	Are we over-reliant on any strategic vendors or partners?
19.	What are your leadership expectations for the CIO?
20.	What would be a stretch objective that you would consider for me as the new CIO in the first year?

This section has posed some questions that you can consider asking the interview panel. Again, you have to watch the body language to see if there is any sensitivity in what you are asking. The questions that you ask are important to probe if you have any direct areas of concern; however, I've always considered that the questions are another opportunity to sell yourself in terms of being a great candidate for the CIO role.

Therefore, I would suggest that you don't ask too sensitive questions in this round. That may be best addressed via HR contacts instead.

The next section explores the interview questions for the CISO.

CISO questions that you may be asked

The CISO interview can be a bit more technical than the CIO interview. However, there are number of questions from the CIO interview that could be repeated – remember that it is likely that the CIO is a key interviewer and decision maker.

Many CIOs will admit in the interview that they are not cybersecurity specialists, but I would not be lured into any false hope that the questions will be easier. The CIO has more to lose if they select the wrong CISO, therefore you should expect a probing and quizzing experience.

Also, in preparation for your actual interview, I would suggest that you write some of your own notes for each of these example questions. Note that these suggested answers are to provide some examples: you should consider how you want to address the question in a way that demonstrates your own position. The actual words you use that reflect your brand are just as important as how you deliver them to the panel.

1.	**What will be your strategic cyber priorities for the next year?**
	Note that you have five key points to talk about for this question:
	• Refer to your 90-day plan where you have a number of areas to focus on initially, including the IT and cyber strategy/roadmap
	• The strategic priorities will be aligned by building an understanding of business plans
	• Stakeholder engagement will also provide some inputs
	• Assessing and rebuilding the team
	• Reference your support for any known strategic or regulatory requirement
	The key point is cyber priorities are subject to new threats that constantly emerge, along with regulatory requirements. This roadmap will require business engagement and commitment for resources/funding.

| 2. | **What is the cyber culture you will be trying to build?** |

Acknowledge that you need to learn what the team has done and where there are gaps. But note how you would want the cyber culture to be based on the following:

- Everyone has the responsibility to protect the company
- Higher-risk roles will be identified, and they will have to take extra care
- We operate collaboratively across business and tech to defend ourselves
- Good behaviors are encouraged and rewarded
- Bad behaviors are discouraged and called out

The key is that cyber culture can always be improved. We want all our staff to be cyber-aware and take on shared responsibility to make our organization more secure.

| 3. | **How will you engage with your key business stakeholders?** |

Refer to your 90 day-plan:

- What are their key issues?
- Where would they like support?
- What would you like cyber to do differently?
- Listen to their responses
- Take action on any quick wins
- Engage them in the cyber strategy review

The key is that the CISO values business engagement and sees this as a critical part of their role. Being seen as a business partner who listens and brings subject matter expertise is important.

4. **What is your top cyber priority?**
This is a hard question, and the answer will vary according to industry and the relative maturity of the enterprise. My suggestion is that you signpost that you would have to take this question on notice and review it during your 90-day plan. Of course, your cyber objective would be to protect the company from the impact of cyberattacks and prepare your defenses for such events.
Initially, you have to understand where we are and where there are gaps, and do the following:
• Assess the current cybersecurity foundations
• Review the control plans to assess our progress
• Consider a red-team test to learn about critical kill-chain gaps
• Review and engage the cyber team
• Engage management and the board by sharing your findings
The key point is you have a five-step plan to assess the foundation, controls, etc., and then set the priorities.
5. **How do you build trust with management and the board?**
Acknowledge that trust will be earned over time with management and the board. The following are a few notes:
• Be honest and transparent (even when this is personally painful)
• Establish security baseline and share this
• Commit to a cyber transformation program to remediate gaps
• Ask management and the board to be role models to build a new cyber risk culture
• Be sure that management and the board understand and practice their role in a crisis
The key point to make as the new CISO is that you will have to earn and keep the trust of management and the board. Trust will be earned over time by the work of the team and your leadership ability to reduce the risk by executing your plans.

6.	**Describe a serious incident response event and how you handled it.**

Select an example that can be shared. Note a few key points:

- What happened and why this incident was concerning

- What did you do to contain the incident

- How did the incident response team act and interact with stakeholders

- How was comms with regulators managed

- How you recovered and restored services

- What you learned

The key point is that you are able to share an incident that was either an area that you were remediating, or one that you can explain without feeling or looking that you failed as a CISO. You want to come across as being process driven and supportive of your team, while sharing appropriately with management and more broadly with an appreciation of sensitivity to externals.

7.	**What cyber architecture changes would you consider?**

This question can be tricky and one that requires caution. My suggestion is that you can't be in a position to decide any changes from the outside. However, the cyber roadmap is to be reviewed in your first 90 days. A few key points that you should consider including in your answer are as follows:

- Need to assess the roadmap and compliance gap

- Compare the architecture direction to the current policies and standards

- Assess the business readiness to adopt *Secure by Design* principles

- Assess the competency of the cyber architect team

The key point is that architecture changes have to be reviewed as part of the cyber strategy review and the roadmap developed after engagement with the business.

8.	**Tell me how you handled a given zero-day.**

Explain a situation where you managed a zero-day successfully. Some notes are as follows:

- What was the example?
- How quickly did you identify and detect it?
- Why was the risk assessment complicated?
- How did you manage communications?
- What mitigations did you consider?
- What did you learn?

The key point is that you provide a good recent example of how this was managed within your team and with business stakeholders.

9.	**How did you address a new regulatory compliance?**

Talk about a recent example of a new regulation. The key points would be:

- Explain the impact of this regulation
- Note how the different teams worked together and your role in leading them
- What specific challenges did this create
- How did you engage the regulator to clarify the requirements?
- What lessons did you learn?

The key point is that you talk about a well-known regulation and how this engagement across a broad group of stakeholders led to a successful conclusion.

10.	**How do you stay current with cyber technology developments?**

Be sure to acknowledge that cyber is a fast-moving domain, hence investing in time to have the CISO and lead team stay current has to be a priority. Note that technology for technology's sake is not the objective, but that one must do the following:

- Monitor significant new shifts and trends
- Stay curious
- Network broadly
- Attend external industry events

The key point is that you stress the ongoing time investment in yourself, the CISO, and with the team as required. Also, it is worth highlighting that external networking is a critical point in staying current.

11. Who do you network with in the CISO community?

This answer may vary depending on whether you have changed locations or sectors. The key points are as follows:

- Broad and diverse networking with peers

- Have the ability to learn from your network and bring the outside view in

- Engage in industry events and forums

- Acknowledge when you have to build this network further

The key point is that you talk about how you network with industry peers to share and gain knowledge. If you have recently moved, then reference this as you discuss that you have some new networks to build.

12. Are you committed to best-of-breed cyber solutions or more integrated approaches?

This is a difficult question and will depend on a number of factors:

- Business financial situation

- Architecture roadmap, noting how many solutions are in the replace phase

- Current inventory and level of complexity

- Current team competency and experience with different tools

The key point is that as the CISO, you will prefer less complexity in the overall technology footprint. Cyber is no exception and it should be a principle to keep things simple but make risk-based decisions that are transparent going forward.

13. How will you engage in the cyber transformation program?

As the CISO this program has to be a key focus. Your engagement should involve the following:

- Strong and active engagement in driving the embedding of new controls

- Accountability for business stakeholder escalations and engagement

- Actively monitor program management and issue resolution

- Sharing lessons learned and success stories with the broader management and the board

The key point is that you must be the active owner of this program. However, there may be some circumstances where this is the regulatory compliance lead. In summary, you have to lead by example and ensure your team is similarly engaged.

14. **What should we do to secure our development lifecycle?**

This is a tricky question, as from my experience many CISOs will avoid getting engaged in this domain. My view is that the CISO has to jointly own this with their digital and engineering leaders. What needs to be done includes the following:

- Understand our baseline
- Gain an understanding of DevSecOps culture in the developer community
- Measure the degree of compliance (code scanning, etc.)
- Review the training approach for the developer community
- Review the tooling
- Measure adoption and metrics to help drive this uptake

The key point is to understand the current maturity level of DevSecOps and what roadmap has been agreed to address these gaps.

15. **What is the hardest cyber challenge that you have faced?**

A tricky question, as you would like to profile your expertise but may be reluctant to talk too much about a serious cyber event that you had to deal with in the past. I would suggest that you do the following:

- Summarize the cyber challenge in business and technical terms
- Detail what you did in the scenario and what information was missing
- Provide a sense of your executive engagement abilities and how you keep stakeholders informed
- Finally, note what you learned personally following this incident, and what you have done since to prevent or detect this sooner

The key point is to highlight your ability to be a problem solver and strong leader. The example that you choose should talk to your personal integrity, resilience, and tenacity.

16.	**What cyber metrics do you care about?**

- You should highlight that this is specifically mentioned in your 90-day plan as a combination of Key Risk Indicators, Key Control Indicators, and Key Performance Indicators that you would want to deep dive on.

  I would suggest that the way to answer is to explain that there are 30–40 metrics that you would be looking to review, and also would need to assess whether these have been measured correctly. State that being data-driven is an attribute of yours and you have a broad set of metrics to monitor. My suggestion is that you would note that this would be addressed in the first 90-day review, however, there are likely to be some themes related to the following:

  - Privileged account access metrics

  - Authentication metrics

  - Vulnerability metrics

  - Third-party supplier metrics

  The key point is that the actual metrics will need further discussion and exploration but there are some commonly used metrics that can be a starting point.

17.	**How do you work with risk management?**

My suggestion is that you would want to highlight how you would seek to partner with risk management. A new CISO will need to have a healthy but robust relationship.

A few key points to note:

- Work in partnership with relevant stakeholders on risk reporting to management and the board

- Collaborate to define the risk appetite and key metrics

- Coordinate efforts to build a risk culture across the business

- Openly communicate incidents, issues, and new plans

- Seek input on tech and cyber control plans

The key point is to talk about how you have effectively worked with risk management in past roles and relate how this synergy improves the overall risk posture by incorporating two different perspectives.

18. **Tell me why data analytics is important to you as the CISO.**

The new CISO has to protect the estate from cyber-attacks. Data analytics is the critical element that is used to develop an understanding of the risk posture. A few examples are as follows:

- Use cyber risk quantification to measure the risk reduction from your investments

- Review threat intel data to see whether any insights can be gained from patterns

- Review security logs to detect threats

- Review of human factors (phishing, education, behavior)

The key point is that as the new CISO, you must be fact driven and use data to drive your actions in terms of where you focus your attention (and where you don't).

19. **What will you do in the first 90 days in the job?**

You can smile as you have this as a written plan and can talk to the interviewers through it.

However, be sure to ask for input, and check whether you have forgotten any key areas of interest or stakeholders to engage. I would never assume that the draft plan would ever not benefit from input from your boss and other key persons.

The key point is that this is a plan, and you will take input from stakeholders as you meet them. But you have some clear areas to focus on to accelerate your orientation and therefore allow you to assume responsibility faster.

20. **How do you balance risk management and regulatory compliance?**

There is some context required here – does the organization already have any overhanging regulatory compliance issues? That is worth getting clarification on. My suggestion is that it is worth noting that your role as CISO is to protect the enterprise, and that means that real risks – meaning the highest-level risks – are your focus.

You may have situations where you can serve both purposes; that is, the regulatory requirement also equals the highest-risk area to be addressed. However, in my experience, these may not be correlated and of equal priority.

As the CISO you have to balance risks and regulatory compliance. Only when there is a trade-off will this get more difficult and senior management and perhaps the board will have to be engaged to help resolve such conflicting priorities.

The key point is that you as an outsider don't have full visibility of the internal weight and degree of past history. It is worth treading carefully on this question. The safe answer is that as CISO, getting this balance right will be an objective.

21. Tell me about a time that you were faced with large budget cuts.

A new CISO may not have had this experience previously but would have been a direct report to another CISO when they dealt with a large budget cut. When there are large budget cuts, then all prior assumptions no longer apply. The key is to evaluate your cyber and IT strategies and understand the impact of these budget cuts on other related projects. A few key points to note:

- Review the current spend and resource allocation

- Assess the impact of stopping projects and operations

- Explore alternative approaches

- Engage the lead team to understand and help navigate the targets

- Communicate with sensitivity to the broader team

- Be decisive

The key point is that you have to be commercially sensitive to business pressures. This is where you underline your focus on the details and your ability to work through tradeoffs. It could be noted that some architecture shifts from best-of-breed may take time to plan and execute.

22. What do you consider to be your weakness?

We have covered this question in the brand discussion. Taking my own example, I would talk about inattention to detail being my weakness and underline my strategic perspective as my strength. I would note that I have been self-aware of this trait for many years and have been working on improving this.

I would give the example of financial management and forcing myself to understand the fine details of the budget. Alternatively, I could discuss evaluating cyber architecture to appreciate the detailed layers of defense and which parts are working effectively. This way, I acknowledge my areas requiring development and can comment on how this is being addressed.

The key point is to be able to show your ability to reflect and act on areas to improve. Clearly, there are some weaknesses that if you discuss may disqualify you immediately, so be mindful of this.

23.	**Tell me about a time that you were stumped and didn't know what to do.**

This question is all about your problem management skills. A few tips are as follows:

- Select an example that was difficult and is also relatable to the interviewer
- Summarize the scenario and what happened
- Highlight how you recognized that you were stumped
- Then explain how you proceeded
- Who did you network with?
- How did you analyze the problem (techniques, approach, etc.)?
- What was the breakthrough moment that helped you solve this problem?
- What doubts did you have?

The key point is that you talk about a time when you were able to overcome an issue and solve the problem.

24.	**What in cyber worries you the most?**

My suggestion is that you acknowledge that as the CISO, there is much that worries you about cyber. But as you have not yet joined the company, there is much to learn about the actual defenses that are in place and those that are being added.

From the earlier interview discussion, you would be able to note some of these:

- Recent cyber activity in the industry
- Nation-state cyber-attacks increasing
- Ransomware attacks
- Supply chain and third-party incidents
- Vulnerability management in light of recent zero-day attacks

The key point is that the CISO must always be on guard and never take their current status for granted. It is likely that an overconfident CISO would not be successful in an interview, except in certain circumstances such as post-cyber breach.

25. **How do you handle stress?**
The role of the CISO is inherently stressful. You should note that being resilient is a necessary part of the position. I would personally talk about how I manage my own mindfulness by practicing yoga (insert your mechanism). A few other notes are as follows: • Focusing on what can be done today with clarity on what can be planned for tomorrow is an important perspective (focus on action) • Note that you have to trust your team to help you defend the enterprise (trust) • Maintaining a good work-life balance is not easy but part of the approach (balance) • Learning mindset (learn from stress, rather than overcoming it) The key point is that the CISO role is naturally difficult, but you will want to be seen as calm and considered with a positive approach to stress management.

In this section, we have explored questions that you can prepare to answer in your interview. Again, you should be prepared for these examples and others that could be asked. Selecting the right CISO is super critical, so the interview will be difficult and designed to challenge you.

Despite the global shortage of cyber talent, there will be a high level of concern on ensuring the CISO can perform the role and uplift the security posture. At this point, you can take a breath and consider what you want to probe. In the next section, we explore the questions that you should consider asking the interviewers in your CISO interview.

CISO interview questions for you to ask

Once again, you have reached the stage in the CISO interview where the interviewer then asks you what questions would you like to ask of them. Having no questions is never going to be seen in a positive light. Conversely, asking too many questions may also be not ideal.

My advice is to ask three to five questions and have these prepared prior to the session. It is fine to reference your notes as this can demonstrate that you are taking this role very seriously and have come prepared. Here are twenty questions for you to consider:

1.	How aligned are management and the board with regard to the current cyber risk posture?
2.	How well is the cyber strategy understood across the business?
3.	How willing are the business units to trade off their system availability for required critical patching?
4.	Does the board and management have the appetite to fund the current remediation gap?
5.	How would you describe the current cyber risk culture?

6.	How do you assess your third-party partners' cyber readiness?
7.	What is the mechanism to stop shadow IT, including shadow APIs?
8.	Are there improvements to the current cyber risk governance process that are required?
9.	Do our subsidiaries have to comply with our cybersecurity approach?
10.	Are we concerned with insider cyber risk? Have we taken measures already?
11.	Is cyber risk quantification understood by management and the board?
12.	What is the brand of the cyber team and how does this need to change?
13.	What changes are required in the cyber transformation program to make this more successful?
14.	Have the cyber strategy and IT strategy been well aligned with the business requirements?
15.	Do we have a large legacy environment that has no plans for renewal?
16.	Has the management team understood and practiced the crisis plan?
17.	What is the largest challenge that we face for business growth/ transformation?
18.	What surprised you about this company when you first joined?
19.	Do we understand which of our most critical vendors have cyber weaknesses in their controls?
20.	What is understood to be our current largest cyber threat?

The interview is over! You can graciously thank the interviewers for their time and for answering your questions. You have learned more about the opportunity and also spent time with some of the key stakeholders. In your mind, you likely have both positive and negative thoughts, but already have decided whether this opportunity is the right one for you.

Summary

Getting through the interview process to be a CIO or CISO takes some preparation. There has to be some considered reflection prior to the interview sessions to ensure that you have a clear mind and are able to put forward the best version of yourself.

The interviews can be considered a barrier to getting a role that can be overcome with the right preparation. Firstly, for the CIO and CISO positions, there will be some probing questions that you have to answer, and there will also be an opportunity to learn more about the job and company when you have time to ask your interviewers some further questions. In summary, the interview process is the heart of the entire hiring process and is generally the unscripted part of the process. How you answer appropriately while maintaining eye contact and with a confident voice can make a huge difference in getting the job.

Preparing your plan for the first 90 days will be an asset that you can use in the interview process. The next two chapters will provide you with an approach to getting this plan ready so that you can be successful.

Part 2:
What to Do in the First 90 Days

In this part, you will take a proactive role to understand what to do before stepping into the new building and how to develop a 90-day plan for those first days. Using the template provided as a starting point, you can evaluate where to spend the time and where not to invest your effort. In the examples provided, we will walk through the key activities that you must undertake when starting a new CIO or CISO role. How you prepare for this entry point will play a massive part in your success in what are extremely challenging assignments.

This part has the following chapters:

- *Chapter 5, CIO – The First 90 Days*
- *Chapter 6, CISO – The First 90 Days*

5
CIO – The First 90 Days

In every journey, there comes the first step, and preparing a plan from this starting point will only increase your chances of success.

> **The fundamental question that is to be addressed in this chapter is as follows:**
> How do I write my plan for the first 90 days as a CIO?

I was a CIO for 15 years at Eli Lilly in five different countries; and now I was moving back to a CIO role after two years at KPMG doing strategic IT consulting. In my previous CIO roles, I always had a plan but never specifically committed to writing it down; instead, I carried it in my mind.

The story I will share commences in a new CIO role that I had held at **Credit Union Australia (CUA)** – a mutual bank that was undergoing a period of unparalleled transformation. Personally, I was attracted to the opportunity as a career *spiral* to move back into financial services. While I had held larger roles at Eli Lilly, the size of my role was never a factor. There was a significant degree of risk in the role, as the bank was completing a wholesale banking transformation of everything. It had all the elements of what I was seeking, and I was consciously thinking two jobs ahead. This CUA CIO role would allow me to return to CIO and then CISO roles in financial services.

The previous CIO (Barry) had recently retired, and he had been in the role for over 18 years. He was respected for what he did, when the bank had been in business-as-usual mode and running very differently. Core banking projects involve a high degree of risk, and many fail to deliver. The transformation change required, however, also included mobile, online, payments, and the entire IT stack.

I was interviewed by Chris Whitehead, the CEO at CUA, and he had previously held CIO roles at St George Bank, now part of Westpac. Clearly, Chris understood business-enabled technology change; however, except for his expertise, there was little going for the transformation to make it a success. Chris had commenced this program of work and signed a contract with Tata Consultancy Services to provide a new core, mobile, and internet banking solution. The IT organization was a traditional shop with IBM mainframe and COBOL programmers who delivered an annual release of their core banking service. CUA had dodged a bullet by moving its mainframes a few years earlier as the water levels rose in the Brisbane floods. Therefore, they had some regulatory scrutiny.

As I settled into my new role as the CIO at CUA, the weight of responsibility pressed upon me. The bank was in the throes of unprecedented transformation, and I found myself at the helm of a ship navigating new uncharted waters. This was an organization that was not used to transformation, and being member-owned, it could not afford failure. And so, it became evident that a written 90-day plan was not a luxury; it was a necessity. The first three months would define my trajectory – building relationships, understanding the business landscape, and aligning strategic goals. Without it, I risked floundering in the tempest of change. In this chapter and the next, we will delve into why aspiring CIOs and CISOs must embrace this discipline, grounding themselves in purpose and direction during those critical early days.

We will cover the following key topics:

- Understanding the need for a 90-day plan
- A brief overview of my first CIO 90-day plan
- Exploring *People* in the 90-day plan
- Exploring *Process* in the 90-day plan
- Exploring *Technology* in the 90-day plan
- Building your plan for the first 90 days
- Asking yourself the hard question

Understanding the need for a 90-day plan

When you start in a new CIO or CISO role, there is always much anticipation. The prior incumbent would have announced their departure, and the team and your future colleagues would be in a state of anticipation – expecting both good and bad to come from the change. There is always going to be some natural anxiety from the technology team regarding work that is in progress. In addition, they might feel a sense of loss in that all the deliveries that had been completed before might perhaps not be recognized.

It is in this spirit that I wrote the 90-day plan. The plan was to both guide myself and inform my stakeholders in what areas I would spend this time. By default, it would also detail where I wouldn't be spending my time, which is also instructive to the broader team. As I noted earlier, I had transitioned through five earlier CIO roles that also had their own first 90 days. This new CIO role was not close to being as large or complex as some other CIO roles that I had successfully navigated. However, I really wanted this to be a winner. Therefore, I formalized my 90-day plan as a written document.

But why is this a must-do for you too? To get offered the CIO role might have been a career ambition for you. And while this is likely not to be your last CIO gig, you would really want to make this first role a success. There will be other larger CIO roles or Group CIO positions that you will want to consider in the future. But if you fail at this initial attempt, then that will certainly reduce your probability of success. Using this approach, I would reference the 90-day plan at my lead team meetings, and then email it to my staff for their consumption and so that they could also share it with their teams. In addition, as I met with all my peers and key stakeholders, I would share the 90-day plan, with one slight variation – I would also ask for their input: "*Have I missed anything?*" This is also an opportunity to elicit any *quick wins* that they would like me to consider. This is specifically called out in my 90-day plan.

Setting this tone will be critical to your success. If you don't set a tone yourself, then you will find others doing it for you, which may not be to your liking. For instance, it would be quite commonplace for your first month's schedule to be full, with internal and external stakeholders all wanting to schedule meetings with you. Generally, I would focus on the internal stakeholders first and only take meetings with external stakeholders regarding issues that were time-critical or regulatory-based.

Having first established your own ground rules with your team, and in particular your executive assistant, decide how you want to work together. There are many things to address, and some can't be decided immediately. Some of these are as follows:

- One-to-one time with the direct manager
- One-to-one time with direct reports
- The team meeting structure and format
- Managing external meeting requests
- One-to-one time with the head of HR
- One-to-one time with the CFO
- One-to-one time with other key internal stakeholders

Note that HR and finance are important foundational stakeholders that you must build a strong working relationship with from day one. It is important to *listen* to these stakeholders more than trying to *sell* to them; it would be too early in the first 90 days to attempt to sell a new vision.

With your executive assistant, you must also agree to protect your schedule and allow yourself some *thinking time* so that you are not overwhelmed with superfluous meetings. One easy guideline is to limit these meetings to a shorter duration, around 20–30 minutes. Any requests for an hour's meeting should be scrutinized and take place only if you agree to it.

Your time is precious, and you are on the clock in terms of making your 90 days a huge success. In that regard, being in sync with your new manager is of critical importance. This means not only securing regular face-to-face meetings but also finding out their preference in terms of what medium to use – phone, text, email, and so on.

It is, of course, extremely important that you take the time to run through this 90-day plan with your new manager and ask for feedback and input. It is likely that they will have already suggested some potential quick wins to consider.

Your own personal test would be to figure out how to assign those potential quick wins to your team and, in some cases, combined tech/business teams to address them. It is likely that the incumbents that you ask will have tried or failed in their attempts to address these in the past. This is where your emotional intelligence has to be preeminent, but also be aware that you will probably have to be hands-on to drive the transformational change and ensure that this is indeed a *quick win*.

Your new manager will have staked their own reputation in bringing you on board, and that is something you have always been aware of. It is not blind loyalty but a realization that this leader has your back and wants you to succeed. Each manager will have a different set of expectations, as they themselves are at different stages of their own career. Some will want to know all the fine details; others will only want to be notified about important strategic aspects. But all managers will want to know about bad news as soon as possible and not have it sprung on them. Yes, you want your boss to look good and, most of all, not make them look bad.

Armed with an understanding of the need for a 90-day plan, let's now shift our focus to the practical steps involved in crafting an effective 90-day plan. In the following section, we'll break down the process, providing actionable insights and guiding principles to help you build a roadmap that accelerates your impact as a technology leader. Let's now take a close look at how you can build your own 90-day plan.

A brief overview of my first CIO 90-day plan

In this section, I will explore the elements that make up a 90-day plan for a new CIO. This section intends to provide some credible examples so that you can prepare your own plan, either by reusing the template or taking it as a starting point. As this plan will also be shared with all stakeholders and staff, they will be aware of where you allocate your time. Here are a few guidelines for the CISO plan:

- The plan must fit on one page
- It must be clear as to what you will focus on and what you won't focus on
- It is designed to be openly shared (avoid any confidential references)
- It must be structured with time-based deliverables (with pressure to stay on track)
- People, Process and Technology will be plotted along the *y-axis* (refer to *Figure 5.1*)
- The plan should be as closed-ended as possible, with more measurable actions at the end of your 90 days.
- Track your progress as the plan is completed (remember that this will be an exercise on branding yourself too)
- Finally, it is not supposed to be perfect, so revise it as necessary

There is a 90-day plan template shared in *Figure 5.1*. Note that your own CIO 90-day plan will have to adjust based on the scenario that you inherit. There will always be an existing agenda that you now own and have to ensure that you understand versus some other areas that may not yet be a priority.

	30 Days	60 Days	90 Days
People	**What does success look like?** Ensure you have clarity of expectations from the CEO and the COO. Get support for this plan.	**Engage executive stakeholders** Share with the CEO and the COO your assessment of the situation.	**Contract success** Ensure specific support is secured.
People	**Understand key players** Do stakeholder analysis and influence map. Ask execs 3 things about their key issues for support and advice. What would you do differently?	**Start to build relationships at the exec level** Understand what their pain points are.	**Build coalitions at the exec level** Look for examples where 'no' is the appropriate answer.
People	**Brand yourself** Set your agenda and expectations–discuss with team, peers, and management your 30–60–90-day plan. Ensure they understand what you value.	**Be visible and allow yourself think time** Set time to wander around. Set time to reflect on ideas with mentors and confidantes.	**Be generous** Give public recognition to your team for leadership behaviors that meet and exceed your expectations.
People	**Assess the team** Evaluate the current IT team.	**Build your future team** Identify talent gaps and take selective decisions to remediate the team.	**Build your team** Start to take action on decisions with your team.
Process	**Understand how IT engages the business** Explore business integration and governance gaps.	**Consider alternatives and options** Strike the right balance (evaluate where rigor and discipline is lacking).	**Take action** Implement deliberate efforts on governance.
Process	**Establish personal key metrics** People respect what you inspect. Ensure this is well known.	**Monitor metrics** Ensure visibility of metrics and be committed to communicate actions.	**Monitor and refine metrics** Give recognition and market the results.
Process	**Accelerate business learning** Ask for access and inclusion. Understand what external models your business partners value.	**Reflect on learning** Be open to external environment and alternative benchmarks.	**Demonstrate learning** Share learnings with the business at meetings and also share these with your team.
Technology	**Fix a critical *hygiene* issue** Identify and address the highest visibility issue.	**Work on the issue** Ensure short term vs long term trade-offs.	**Ensure accountability** Trust but verify.
Technology	**Review IT strategy and strategic projects** Own this. Explain intent, scope, and rationale. Evaluate the portfolio.	**Challenge approach** Understand where changes may be required. Challenge what looks to be unnecessary.	**Get support from the COO for any adjustments** Ensure airtime is secured.
Technology	**Understand Ops and Security** Baseline metrics.	**Improve and refine goals** Measure and communicate metric adjustments.	**Measure and follow up** Show progress–war of inches.
Technology	**Send the right message** Look for a quick win–something to stop.	**Stop something** Be decisive.	**Start the loop again** Share this and repeat with the next issue.

Figure 5.1 – David's CIO 90-day plan

At a high level, note that there are more *People* and *Technology* areas of focus than *Process*. There will always be, of course, personal trade-offs, but it is interesting to consider for yourself whether you agree with this weighting. I'm going to now explore the *People* aspects of this plan.

Exploring *People* in the 90-day plan

There are strong expectations for the CIO to be a leader first and a technologist second. And when I say this, I recognize that the technologist role is a close second, as the CIO has to work across different emerging technologies and the value proposition that this can bring to the board and management. But leading through transformation change is the most critical ingredient.

The CIO must operate effectively in the *People* dimension. I've worked with and seen many naturally shy CIOs who have learned during their careers to be able to overcome their natural tendencies, becoming more outspoken and communicative than they are comfortable with. Being a people person is clearly not mandatory for the new CIO, but their job becomes much harder if they are unable to overcome any natural tendency to be less communicative. During the selection process, there would have already been some testing and consideration of how this new CIO would work with other incumbent stakeholders. That is a natural process that the interviewer will cross-reference with your personal style, perceiving how this would fit in with your fellow colleagues.

What does success look like?

For the CIO, their career success comes from being able to use technology effectively to grow their business. The executive stakeholders need systems to be fully operational and create positive customer experiences, which ensures both stability as well as security.

The CIO must be aligned with the business strategy and be able to demonstrate that the IT team fully comprehends and is aligned with these objectives. Accordingly, the CIO has to become a role model through their actions by being a great listener and being able to transform a business requirement into a technology solution that delivers the outcomes that are expected. This is easier said than done.

In the 90-day plan, the CIO has to use this time to get a level set from stakeholders of what they expect and, indeed, what is unreasonable. This may not always be obvious on day one, and the CIO will face many stakeholders who want to test them and their resolve. Therefore, understanding these key players is critical.

Understanding key players

The CIO will have a long list of meetings that have been pre-arranged before their first day on the job. It is interesting to consider how to calibrate and measure these players on whether they have urgent or critical requests. There will be many management colleagues who will await with a long list of concerns and perhaps issues to address.

My sense is that I always prefer to have visibility of these issues rather than them not surfacing. As the new CIO, there will be meet-and-greet sessions, and you will probably have to start to set new expectations from that very first meeting. You have to acknowledge and take notes from these meetings and commit to investigating the issues identified. However, also ask the stakeholder that you meet to help provide some sense of their own priority and why that would be important to tackle.

There will also be certain expectations coming from your direct boss. This may be the CEO, COO, or CFO. In every organization, there are different reporting lines, but what won't be different is that every manager has a personal hotlist of burning issues. Your challenge would be to quickly sort through and reset expectations if any of these items look too daunting to be addressed quickly. What you have to remember is that you can't make progress on these items alone. While you often won't have a good measure of how strong your new team is, there should always be an expectation that they will deliver for you.

As the leader, you are expected to make an immediate positive impact. It is not always easy, especially when you don't often know who to trust and who can be relied on when pressure starts to build. As the honeymoon period will be over pretty quickly, you have to make personal judgments and be able to judge the personal competencies of your team and identify the gaps you have to address.

Taking charge and being seen as the leader will always be required. What you have to also realize is that you have to do it in a manner that honors the past and is respectful. It is always best not to criticize the previous leader and what was achieved or not achieved. While it may be true that the previous incumbent had not been successful, you risk putting off some of the staff, who may then feel slighted.

Looking forward in terms of what the challenge on the table is and how you want the team to come on this journey with you is the right approach. This will, from my own experience, always resonate with the team. Remember that the team must have a strong sense of belonging, and the leadership team creates a culture that defines how the team delivers on its goals.

Key action points

- Engage in pre-arranged meetings with various stakeholders, and seek clarity on stakeholder priorities and urgency levels
- Prioritize and, if necessary, reset expectations from your direct boss considering a team's capacity
- Quickly evaluate your team's capabilities, and make informed judgments about team members' competencies
- Take charge and be seen as a leader who defines a positive vision for the team's future journey. This will foster a sense of belonging within the team

We live in a world full of brands, and leaders are no different. In the next section, we explore how you create a brand for your CIO role. These words make it sound like a marketing and promotion operation; however, that is far from the truth, as it's more related to personal branding, which we all have. The question is, does this branding help or hurt you?

Creating your brand

As you walk into the new CIO role, there will be ideas and perceptions of your brand even before you start. There might be someone who has heard of you or knows someone that you worked with previously. There will be a brand that accompanies you when you arrive.

As a leader, you should have personal confidence that your career in this new role will be a total success and will lead to more challenging and, perhaps, larger roles. But that is not what matters now.

You must create your own brand and understand what will be valued and expected of you. I've always personally believed that your brand is *not* what you say about yourself. Your brand instead is what people (staff and stakeholders) say about you when you are not in the room. The staff will watch and observe whether you act consistently in your words and actions. They will also be checking whether you act differently with different audiences. In each scenario, they are judging you, and your own actions will build or destroy trust.

As a leader in these first 90 days, you will set your expectations as to what *normal* means to you. If you are driving a change agenda (which is likely), then you would need to speak about change and demonstrate your commitment toward it. The best leaders will strive to be **servant leaders**, and the foundations for this can be set early. Your objective should be strongly *people-focused*. Having the right blend of open- and closed-door discussions is key. I've seen many leaders take on too much at either end of the spectrum, or not engage broadly enough at the town halls or skip level (i.e., talking to the staff of your staff). Being people-focused is the brand that you should strive for, and this will help you start to build the natural support in the team that will be required to make your plans possible.

Key action points
- Understand that perceptions of your brand exist even before you officially start
- Acknowledge the existing brand and consider how it aligns with your goals
- Strive to be a servant leader, focusing on people and their needs
- Cultivate a brand that encourages natural team support

Engaging others may sound too obvious to mention, but I know that there will always be a temptation for leaders to get straight into problem-solving mode in a new role. While this is required, it is essential that there is time set aside for engagement with your new peers and staff.

Engaging your peers and staff

As the new CIO, you will be engaging broadly across an organization; it will feel like a roadshow. There will be faces and names that you will automatically remember and recall. Plus, there will be many things that you will struggle with. My tip for this stage is to ask your executive assistant to prepare pages with names, titles, and photos that you take notes with for future follow-ups.

To lose track of who asked or suggested what is not only embarrassing but will also give you the negative brand of not listening well. There will be some stakeholders and staff members that you may not want to wait to follow up with when they will bring a new pressing item to you. This is to be expected.

I've made it a point to try to help my staff with some of their personal concerns by being as transparent as possible as to what I care about and what I expect of my team. In addition to sharing the 90-day plan, I have presented and shared a simple "My Expectations" page with some 5–6 bullet points.

This may be slightly different from what you would want to share, but here is what I have used in some form:

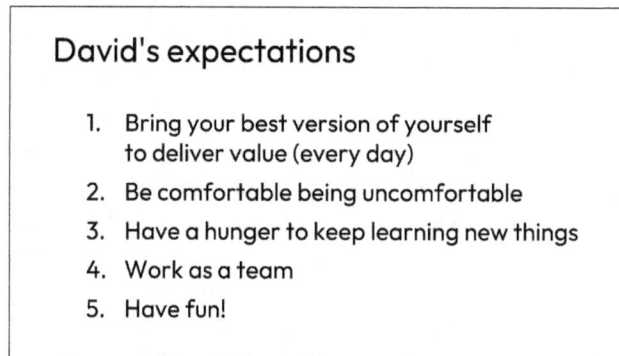

```
David's expectations

  1.  Bring your best version of yourself
      to deliver value (every day)
  2.  Be comfortable being uncomfortable
  3.  Have a hunger to keep learning new things
  4.  Work as a team
  5.  Have fun!
```

Figure 5.2 – The David's expectations slide for town hall meetings

This shortlist will raise some questions, and your staff will want clarifications about what you mean by being uncomfortable. It is meant to be simple, but, at the same time, difficult. I've seen this simple list, which I have used in many C-suite roles, pinned up by my team members as a reminder. It's not to flatter me; it's just human nature that people want to understand what is expected and test themselves to see whether they agree or can do something.

Here are some other useful tips on how to proceed once you have completed the roadshow to meet all stakeholders and your staff:

- Taking the time to test your memory of people's names is a good activity, and it will build your own brand as a person who cares enough to try to learn about their team.
- The observations that you have gathered should be shared and tested with your own leadership team, and you want to see their responses to see whether you have made the right analysis or

not, at the right time. This is likely to take place at your team's offsite meeting. Following up with a formal stakeholder analysis and influence map is important (see *Chapter 6* for examples).

- You have to build bridges (coalitions) where they don't exist and strengthen these relationships where they have been built. There may be some baggage that comes along with your team, and the first few days would be a great time to remove this baggage in cooperation with key stakeholders. I would suggest that you are as direct as possible, and although you are at the early stages of your relationship with the team, a smart approach to take would be to ask what they think needs to change. This doesn't mean that their request will automatically be fixed or, indeed, is fixable. What you don't want is these hidden agendas to affect your relationship with a key stakeholder.

- Once you have broken the ice with them, then you should engage them to help you get up to speed on their business – to understand their strategy, what really matters, and how your team can best support you. These are just some of the questions to be discussed. In truth, just asking how to learn more and who to contact will help you build bridges with the stakeholders.

Key action points
- Engage broadly across an organization, treating it like a roadshow
- Be transparent about your expectations for the team
- Consider formal stakeholder analysis and influence mapping, and share your observations with your leadership team for validation
- Engage stakeholders to understand their business strategy and identify what truly matters to them. Determine how your team can best support their goals

---

Your new CIO team will eagerly await your arrival. They will be both excited and a bit wary of the new boss. This next section explores how to assess and then build the new team around you.

Assessing and building a team

The new CIO has a honeymoon period that they must take advantage of, and that is the time to assess what talent has been inherited in the role and what gaps should be addressed.

Whenever you start a new CIO role, there will be a natural tendency to reflect on your interactions with the team members and make mental notes of who impressed you and who didn't impress you. That said, you will get some input from your boss and stakeholders in the orientation process, and some unsolicited feedback will be received.

However, it is imperative that you make your own deliberation and decision on staffing matters. The staff will be aware that you are doing this analysis, and it goes without saying that, as the new CIO, you will look to make changes from your predecessor. Your staffing assessment must be cross-referenced with the needs of the IT strategy and operations teams. The IT team's capability and capacity will be largely determined by the scope and complexity of the challenge you face.

The demands of your new CIO role will require you to be considerate; however, you also must understand that the strength of the future team will very much determine what you can deliver or not deliver. Simply put, if you build a strong new team, then your own expectations in terms of business and technology outcomes will rise. Thus, you will establish a new baseline of what good looks like in terms of performance and talent. There must be consideration of the new team's DNA and identifying existing talent that can step up into a more multifaceted position. A new team culture can only be set properly once there is a new future team selected and put in place.

Therefore, this starts with your own assessment, ensuring that your HR partners and associated external search firms understand the new requirements. To be absolutely fair with the process, you have to be transparent and let your team know what you are looking for. This then gives them a chance to reassess their individual positions and perhaps also decide to take a voluntary exit.

As the new CIO, you will have to understand your own strengths and weaknesses, as we discussed in the earlier chapters, and then assess what is needed now to improve the chances of success. This becomes your wish list of **skills, knowledge, experience, and behavior (SKEB)**, which you will have to summarize to work with HR partners and stakeholders in attracting new talent.

Because the role of the CIO is broad, the requirements may also be extensive, requiring a multitude of SKEB analyses that can identify specific gaps that need to be addressed. I have always prioritized this activity in new roles, and this means a significant investment of time spent reviewing candidates' CVs to screen out the best ones, and then engaging them to ensure that you sell the vision and how they will be a key part of the puzzle. This engagement of the new potential staff is essential so that there is a clear understanding of your high expectations of them and their contribution to the mission.

The other thing I would also often drop into selective conversations was that while I was new in the role, I was not intending to be in it forever. I would outline that my mode of operation had been delivering transformation change, and then I usually transitioned to the next challenge in three to four years. This served to motivate those who were ambitious to develop themselves to take over my role. Again, this reinforced the alignment of my own intentions and objectives with members of the team.

Key action points
- Reflect on interactions with team members, take mental notes, and consider input from bosses and stakeholders while assessing existing team members
- Make your own deliberations and decisions on staffing matters, and align your assessment with the needs of IT strategy and operations teams
- Establish a new baseline for performance and talent, and identify existing talent that can step up into broader roles
- Properly reestablish a team culture once a new future team is selected

In this section, we have explored the *People* aspects of the CIO 90-day plan. This is a critical focus area and will enable your success when you get this right. Next, I will discuss *Process* aspects and why these are important for your new role.

Exploring *Process* in the 90-day plan

As the new CIO, you will be inundated with invitations to different governance meetings – risk committees, management committees, steering committees, policy committees, and so on. At first, this can be particularly useful to meet and work with your stakeholders on common agenda items, enabling you to understand how they think and what they care about. However, this will likely become frustrating, as you will want to spend time on what you consider critical deliverables in your own 90-day plan but not get the time to do this.

Indeed, you will have to learn to understand that *Process* can both help and hurt your cause. It requires an orderly process to have strategic projects prioritized and resources allocated. In the absence of such a facility, you will become branded as the bottleneck. There also needs to be an operational process that requires you to agree on systems availability and testing. All these activities have potential customer experience impacts for the stakeholders, and you will want to manage the vulnerability and obsolescence risk. Again, these processes help you balance the business and technology risks in partnership with key stakeholders.

All of the preceding help the new CIO become orientated with a business. There must also be a specific focus in the 90-day plan on understanding how IT engages the business, which I will discuss next.

Understanding how IT engages a business

The CIO will spend much of their early days with the heads of business, but it is likely that the IT team will request a lot of the CIO's time which they might not be able to give them. Since this does send the right message to the IT team, I have been transparent to the IT team to let them know that this may be the case.

Thus, you become a role model for the behavior that you expect to see from others. The key is that you share with the IT lead team what you see and hear on your governance process rounds. You would want to test these with the team to understand whether there are blind spots or whether certain issues have previously been raised but not acknowledged. Having the team themselves engage in this discussion is healthy and points everyone in the right direction.

What you want to be able to achieve is understanding where there needs to be more intensive or less intensive engagement with stakeholders. The priority of this engagement must be based on the positions these stakeholders take in your strategy. However, this priority may also be required because a person has an urgent need, or they can be an ally with the power to help your cause.

The first focus is to understand and assess the degree of business engagement in the IT team's strategy and how this process works across different business units, products, and geographies. Ideally, your IT team has a seat at the table and is integrated well into the process, ensuring that IT is well aligned in the business. This is often not the case, and the CIO must cultivate this expectation with their management peers and the team. Note that this may start a bit clumsy, as you are simultaneously reviewing your future team. Therefore, there may be some reassignments in the team, but my advice is to start and not wait, as this may delay you by three to eight months while new roles are filled.

Key action points

- Be transparent with the IT team about your engagement with the business. Let them know that you may be spending more time with business leaders. This helps manage expectations

- Regularly share insights from your interactions with business stakeholders during governance process rounds. Discuss any blind spots or issues raised by the team

- Assess how business engagement works across different units, products, and geographies. Ideally, ensure that IT has a seat at the table during strategic discussions

- Begin engagement efforts promptly, even if it means dealing with team reassignments. Waiting may delay progress by several months

---

How a business is engaged is a great segue into what I would like to discuss next. In your first 90 days, there will be some metrics that you hear quoted by your stakeholders that they are concerned about. Conversely, as the new CIO, you will create a brand around what personal metrics you care about, so this is critical.

Establishing personal key metrics

There are many kinds of metrics that a CIO will need to understand. This includes the following:

- **Key Performance Indicators (KPIs)**

- **Key Control Indicators (KCIs)**

- **Key Risk Indicators (KRIs)**

Each of these serves a purpose in providing the new CIO with specific insights. For instance, the KPIs can provide a measurable metric on team performance over time versus a specified target. KPIs can also be related to on-time project delivery. In truth, there are thousands of combinations of KPIs that are possible. Let me discuss a few of the most important KPIs relevant to a new CIO to review:

- **IT investment management**:

  - **Total cost of ownership**: What value is delivered for all hardware, software, staff, and other costs, versus the costs being managed

- **Return on investment**: Has the expected **Return on Investment (ROI)** been achieved, such as cost savings, business productivity, or enablement?

- **Benefit realization**: Metrics to track the progress in achieving the delivery of the expected beneficial outcomes from the business case

- **Business alignment**: IT should support the overall business goals – what percentage of IT budget is spent on growing the business versus running it

- **System performance and availability**:

  - **System availability and downtime**: This tracks system uptime to measure business continuity

  - **System uptime and response time**: This measures the responsiveness and performance of key IT systems

- **Customer satisfaction**:

  - **Deployment success**: This tracks the successful deployment of new technologies

  - **Net Promoter Score**: The **Net Promoter Score (NPS)** measures user satisfaction with IT support

- **Cybersecurity and compliance**:

  - **Cybersecurity incidents**: This is the number and severity of security breaches and what impact this has caused on a business

  - **Vulnerability management**: The number of open critical vulnerabilities

  - **Privileged access metrics**: These are a series of measures for users that have elevated additional system access

  - **Compliance**: The percentage of compliance with mandatory regulatory requirements

- **Innovation and project management**:

  - **Project delivery**: Tracking the overall portfolio delivery status

  - **Strategic innovation**: Tracking the progress of digital transformation initiatives into production from the pilot stage

  - **Change management**: Measuring the effectiveness of organizational change aspects of large IT projects

This list is just a sample of the metrics that a new CIO can consider, and there are many other potential metrics you will want to review. This is part of your deliberate consideration, as you want to not only understand and learn but also avoid creating additional work unless you believe it to be of longer-term value. The creation of new metrics will cause some consternation and work for your team. While this sends a clear message on where you must focus, there will be times when you would be advised to understand the whole end-to-end process. And you don't want the team to invest time creating new metrics. So, when asking for new metrics, make sure to be clear what they are, and only ask for them when you really need them. I have seen new leaders make these requests and the team become disappointed later when there appears to be no outcome from this effort.

The old saying that the *staff will respect what you inspect* is very true. But they also will lose respect if you don't follow through to act on what you have asked for.

There are also KCIs that the CIO will want to review in detail. These should already be in place for the critical controls. The KCI will provide a measure of the current state of effectiveness, hence some risks that may be important.

As the new CIO, you will want to ensure that there is good control governance so that the homework of the IT team to improve control weaknesses is defined and progresses as per the plan. If you don't see this in good shape, then it would be a high priority for you to insist that this be prioritized. In truth, the safety of your own job is impacted by this.

Similarly, the KRIs, which are there to measure the *risk appetite statement*, should also exist and be in progress. These KRI metrics tell the board if the enterprise operates within or outside of risk tolerance. Hence, they are critical for you as the new CIO. You will want to ensure that these KRI metrics are fit for purpose in terms of providing a view of the residual risk summary, without trying to measure everything. These KRIs will be metrics that the CIO will report to the board and management at various risk committees, and you will be held accountable for driving improvements that need to be made. Therefore, if you see any concerning KRI metrics, then it is best to speak up in the first 90 days.

Key action points

- Familiarize yourself with KPIs, KCIs, and KRIs. These metrics serve specific purposes and provide insights into team performance, project delivery, and risk management

- Request new metrics thoughtfully. Consider the value they will provide and avoid unnecessary work for your team. Balance curiosity with practicality – don't ask for metrics unless you genuinely need them

- KRI metrics are reported to the board and management in risk committees to measure risk appetite; hence, they get significant attention

- Hold yourself accountable for driving improvement based on these metrics

This is all part of accelerating your business learning, which I will discuss in more detail in the next section.

Accelerating business learning

There is still much to learn about a business to understand what it does and what it really values. You want to get a seat at the business leaders' tables to be involved in the discussions and be a true partner, not an order taker.

Once you have access and are included in discussions by default, then you can make a real difference as the CIO. Unfortunately, this does not always happen straight away for the CIO, and there may have been reasons why your predecessor was excluded from discussions. It is always best to make it clear to your peers and boss that this is how you want to operate and that this expectation needs to flow also to your lead IT team.

By gaining proactive engagement, the new CIO can build a deeper understanding of the business strategy and how technology supports and enables this. There are nuances to different business units that are beyond the standard jargon, and it is fine to admit to fellow peers that you might not understand certain things. This context will help you understand the business strategy better and, hence, allow you to build an IT strategy that is more aligned with the business.

An example of this is to gain an understanding of the end-to-end business processes used by the front-line sales and marketing operations teams. This may involve spending time in the field to really grasp how the technology supports or does not support the teams' missions. I've personally spent time in branches, hospitals, concrete plants, and various low-tech environments.

Given that often your mission is to digitize these processes, this experience will provide insights that you can't get from a flow chart or online interview. As you reflect on what you have learned overall, then, as the new CIO, you can take immediate action and have an impact now, rather than waiting for the first 90 days to be completed.

I always believe that a good leader is never afraid and shows courage to make important decisions. There will be projects or activities that you will look at and immediately be puzzled as to why they are part of your portfolio. To then engage the team to decide early to *stop* or *continue* what is already in progress is a hard but necessary call.

You must make hard choices, as this defines a strategy. As a result, there will be strong echoes in the hallways of the organization that you have stopped a project or activity. This will be either applauded or create concern. The strategy will be refined, and the impact you make should free up time for a more strategically important task. This is indeed a win or, at least, the beginning of a potential quick win.

How you spend your time will determine how much impact you can make. From my experience, you must aim for a status quo balance that looks something like the following:

- 30% of your time with direct reports and coaching staff

- 20% networking with peers

- 25% on transformation projects

- 15% in operational meetings

- 10% with external customers and learning best practices.

How you should spend your time

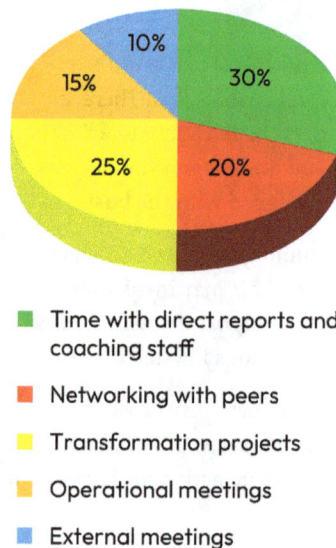

Figure 5.3 – Where should a CIO spend their time?

There is no wrong or right answer here, but I did work with some great **Leadership Development** (**LD**) professionals who confidentially measured the time spent by each member of management, and the CEO shared this at a leadership meeting (without any prior warning). The intention was to get the right balance of discussion and encourage the management to be curious, rather than always advocate a position. At the end of each meeting, the summary was shared, and this drove behavior change around the degree of participation and how you did this. The CEO repeated this exercise for another three to four months, and without telling us what the right time percentages were, it was clear that most leaders had to adapt and change to have the desired impact. This was a great life lesson.

> **Key action points**
> - Aim to be involved in business discussions rather than being an order taker. Make it clear to peers and superiors that proactive engagement is your preferred approach
> - Gain insights by spending time in the field and observing front-line sales, marketing, and operational teams
> - Take action based on what you have learned, even before the first 90 days are complete. Be willing to stop or start initiatives based on strategic alignment
> - Allocate time strategically. Free up time for more strategically important tasks

Now, let's move on to the Technology aspect of the first 90-day plan.

Exploring *Technology* in the 90-day plan

It seems silly to say it, but technology is fundamental to the role of the CIO. This is the core element to ensure that the IT strategy is comprehensive and integrated into the business objectives. The CIO must work with the business and drive innovation in their digital journey. And in doing this, the CIO must be articulate and able to translate tech jargon into business value.

The CIO must spend time building their understanding of the current IT environment and knowing what the plans are. This doesn't mean that this is an action taken just with the CIO's direct reports; instead, it requires extensive business engagement.

Reviewing the IT strategy and strategic projects

In the first 90 days, the CIO will want to have a deep and hard look at the IT strategy and assess whether this document is fit for purpose. Having an approach that involves making choices of what to do and what not to do is the essence of a good strategy. If there were no trade-offs, then I would be very concerned.

The strategic roadmap will be what you are judged upon long after you have finished in your role. However, as the new CIO, you will always inherit the IT strategy that is already in place. Regardless of whether it is suitable or not, there will be some pressure on the CIO to not overhaul the entire plan.

As you step into various steering committees, you will gain a sense of the level of executive sponsorship and how strong a commitment this is. I was always careful to judge leaders by their actions and not just their words, because when there are burning issues within a complex program, some leaders will step up to the challenge while others will disappear.

Your own boss will be able to provide you with some context of the various stakeholders and the degree of their *skin in the game*. That being said, as the new CIO, it is during these first 90 days that you can challenge the approach and things that are already in motion. Yes, as the CIO, you can always provide a level of mature challenge, but the difference is that when you take this position earlier, you may have a better opportunity to be heard.

Addressing any issues with the IT strategy privately with your boss may be the right approach before you openly challenge it in meetings. However, this all depends on the situation, and if you feel that some harm will come from an action, then it is always best to speak up first.

IT strategy is a very broad term and can also include a digital, data, or AI strategy. This all depends on the scope of the CIO role, but regardless of whether these elements are within your remit, the strategy has to be integrated to ensure that the overall strategic impact is considered. For example, in every business that I have worked in, there is a clear bandwidth limit that can be stretched to a degree (I guess you can think of it as a *rubber band*), but there is always a breaking point.

In some cases, the CIO will also have to engage with a chief digital officer, chief data officer, chief transformation officer, or chief AI officer. They may be peers or direct reports, but that does not matter. For the business, they just want the best results, and having an IT strategy that looks at the big picture and is customer-centric will get the best outcome.

> **Key action points**
> - Focus on delivering the best result for a business. Use this approach to make strategic choices about what to prioritize and what not to pursue
> - Acknowledge that you'll inherit an existing strategy but may need to adjust. Recognize that the strategic roadmap will be a key measure of your success as a CIO
> - Use your position as the new CIO to question existing approaches and decisions. Provide mature challenges where necessary, considering the potential impact
> - Understand that the IT strategy may encompass a digital, data, or AI strategy. Ensure integration across these areas for maximum strategic impact

While you are working through the complexity of the IT strategy and perhaps refreshing parts of this document, you also will become aware of *hygiene* issues that frustrate your own team and are prevalent within the business units. These can present great opportunities for you as the new CIO.

Fixing critical *hygiene* issues

As the new CIO, you must identify any current practice that serves to undermine the IT infrastructure. This issue may be understood but perhaps already risk-accepted as being *too hard* or *not that critical* to deal with. The benefit of fresh eyes and a new perspective is that you have made this observation yourself or heard about a practice from a colleague/staff member.

It is highly likely that you will come across several practices that could be considered critical *hygiene* issues. Some good examples are as follows:

- **Shadow IT**: There might be many **Software as a Service** (**SaaS**) applications in use that are unapproved by the CIO, including sensitive personal information

- **A failed project**: There might be an existing project that has failed, but the team is unwilling to share this bad news

- **Access management**: The existing exemptions for shared passwords, admin privilege, password expiry, and so on may not always be the best practice

- **Legacy systems**: There might be many old unsupported systems that are in service, with no plans to replace or retire them in place

- **A cyber culture**: Staff may not be trained appropriately to tackle issues

- **Backup**: The backup process may not be robust in terms of isolation for ransomware protection, or it may not have been tested appropriately

- **Vulnerability management**: There might be strong resistance to upset business stakeholders to patch systems that have highly rated vulnerabilities

- **Authentication**: There might be a reluctance to activate strong MFA, as it may be expected to negatively impact the customer experience.

These are all real examples that I have seen in my career (don't laugh). It can be a real challenge to try to address one or more of these issues, as it could lead to significant friction. However, as the new CIO, you will need to identify that there are a few hygiene issues and, when notifying management about them, take a strong position to insist that the highest-risk issues are resolved immediately.

This will send a very strong message and establish your brand. To tackle any of these takes courage and conviction. So, you will want to double-check your facts before taking the plunge. However, once you take this step forward, you will start to drive strong accountability across your team and be acknowledged among your peers for tackling a challenge that no one else was willing to.

Key action points

- Identify all unapproved SaaS applications currently in use. Develop a process to approve and monitor new applications

- Investigate any existing projects that have failed. Communicate openly with stakeholders about the project status and lessons learned

- Identify outdated and unsupported systems. Develop a plan to replace or retire these systems

- Prioritize patching for systems with critical vulnerabilities. Involve business stakeholders in the decision-making process

---

As the new CIO, you must dig deep to understand the IT portfolio that is in place. You now own this, even though it was not bought, built, or approved by yourself. There is a requirement to get an overall picture of the IT operations, and this includes cybersecurity.

Understanding Ops and Security

This will be one of the many things that you will request an orientation for. The approach will vary, depending on your team, but you want to gain a foundational understanding of the daily operations of the IT infrastructure. The team will start to show you architecture diagrams to give you the bigger picture of the IT landscape. There will be many questions that you will want to ask; let me suggest a few to get you started:

- **The IT landscape:**

  - What are the existing IT policies and standards (i.e., change management, security, and data backup) and are they enforced?

  - What is the current state of the IT infrastructure (in terms of the age of the hardware/software, capacity, and redundancy)?

  - How is IT currently organized? (i.e., the team structure in business versus central team)

  - How many applications are not IT-standard (shadow IT, models, etc.)?

- **The IT reputation:**

  - How well do current IT systems and applications meet the needs of different user groups (by business units, products, and geographies)?

  - What are the significant IT-related challenges faced by users?

  - How can IT better collaborate with others to drive business innovation?

  - Is the business considered to be slow at adapting to new changes, and are there emerging technologies and trends that could help?

- **Current IT performance:**

  - What are the biggest pain points for IT operations (e.g., managing vulnerabilities, skills gaps, and legacy tech)?

  - How does IT measure success (i.e., how does it align with business goals)?

  - What is the IT department's budget, and is this sufficient? If not, why?

  - What existing strategic IT projects are underway, and what are their objectives/timelines? Does it make sense to continue them?

By exploring these questions, as the new CIO, you will gain a foundation to understand the strengths and weaknesses of the existing IT operations. On this basis, the new CIO can look to make improvements and reset the metrics to ensure that operations continuously improve.

Key action points
- Identify any gaps in existing IT policies and standards
- Assess the current state of IT infrastructure, and determine the number of non-standard applications
- Evaluate how well current IT systems meet user needs across different groups, and identify significant IT-related challenges faced by users
- Identify pain points in IT operations, and define success metrics for IT

How well your first 90 days go will often determine whether your new role will be a success or not. Therefore, reflecting on what worked and what did not work as expected is important. You must send the right message in the first 90 days.

Sending the right message

To send the right message as the new CIO, you must have a bias for action but not be guilty of taking on too many new projects and initiatives. It will be tempting to do this to try to prove yourself. However, your own team is being rebuilt, and there is a danger that you could outstretch your capacity in the short term.

Another approach that I would recommend considering is to stop something in your first 90 days. This may be a small symbolic act rather than a major project. For instance, there may be a timesheet system that was put into place in which no one really uses the data, or there may perhaps be reports that you generated that are not reviewed.

I recall a CIO role where I discovered that there were 3,000 reports generated each month that, we noted in the logs, were never opened. That was an easy issue to remedy, even though the team that published this had felt horrified at first.

What I noted was that following this quick win, there was a sense of feeling liberated, and that it was fine to challenge the status quo. You must start building this sort of culture with small steps, taking your team in the right direction so that they begin to demonstrate the right behaviors.

Therefore, it is not worth attempting to stop something that is quite embedded in a system and has become a ritual. These issues are best kept on the back burner to tackle some other day.

Key action points
- Maintain a bias for action, but avoid taking on too many new projects
- Recognize the risk of overstretching your capacity, especially when rebuilding your team
- Focus on quick wins that challenge the status quo, using small steps to build a culture of positive change
- Encourage your team to demonstrate the right behaviors

In the next section, we will look at reflecting on your 90-day plan and evaluating whether it is the right fit for you.

Building your plan for the first 90 days

In 2014, I wrote a story that was published by CIO.com, entitled *My First 90 Days as a CIO*. This was a plan that I wrote before starting the new role, and I had anticipated what I would need to do in these vital first 90 days on the job. The plan was not private, and I used this as an approach to start branding myself. The lessons included my own personal experiences and insights from my previous 20 years as a CIO in different roles globally and in diverse industries. I have tried to list some of those lessons in the previous sections. The basic structure and skeleton of the 90-day plan have withstood the test of time. However, there will always be some minor tweaks required in different roles.

Having a written plan comes with a caveat, as you don't know all the old issues in a new organization. However, in an interview, you would differentiate yourself from other candidates by providing assurance that you have completed some due diligence. Getting the balance right across *People*, *Process*, and *Technology* is always difficult. I would always share my written plan with all the stakeholders, including management and my own team. Transitions are always a tricky time, and not writing down any explicit objectives can be seen as an excuse for being unfocused.

As the new CIO, you will find that there are many other agendas that will get in the way of you doing your job. Everybody has their own priority, and it is not hard to get sidetracked into giving some people and tasks undue priority in your plan. There have been events that have triggered adjustments to the plan, and I've taken the disciplined approach to update it in these instances, rather than freestyling it.

At this stage, you have met all the key stakeholders and understand the job you signed up for. You are still fairly new as a CIO, and while you are not yet fully comfortable in the role, you now have a fair idea of how it is playing out. What comes next?

As you delve into the critical phase of the first 90 days, it's essential to reflect on whether the role you've stepped into aligns with your expectations. You should take some time to evaluate whether this is indeed the career path you envisioned. Let's explore this pivotal question and guide you toward making informed decisions as you continue your journey as a new CIO.

Asking yourself the hard question

It's now time to reflect and ask yourself the hard question, *"Do I like this job?"*

This will be a question that you probably would have been contemplating each day of these first 90 days. No doubt, there are going to be things that you are concerned about, and other items that have been a pleasant surprise. However, it is likely that the job is very clear, and all its intricacies that were never shared during the interview process will now be somewhat clearer.

It is not a question at this point whether you're going to stay or not. Honestly, I've never personally been in a situation that looked so poor that I wanted to exit early. I should add that to quit a role at such an early stage would suggest that you rushed into getting an interview, and you really didn't know what you were getting into.

Purely from an angle of integrity, you should embrace the challenge of the new job and reflect on how to address any concerns you have about it. Unfortunately, it can also be seen as *negative* for your own career track record when there is any history of quitting jobs prematurely.

But what happens when you really don't feel that this is the right place or job for you?

What are the show stoppers?

You need to identify and address the really major concerns that affect your confidence in taking on this CIO role.

Do you know the real reasons for wanting to leave? Here are some questions that you need to ask yourself:

- Is the work culture a poor fit and not what you expected?
- Is the manager who supervises you not what you expected?
- Does the role as advertised not match reality?

If you answer *yes* to any of these, then how can you address the issues?

- Have you tried to discuss the issues one-on-one with your direct supervisor?
- Could you (with sensitivity) test out your concerns with HR?
- Do you have a trusted advisor or mentor who can listen and provide you with a different perspective?

Once you have explored all the possible solutions, make a final decision to stay or leave. It is a difficult one, as your initial enthusiasm and energy may be diminished by any concerns that you have since discovered.

Rinse and repeat

I would recommend that you, in effect, redraft your 90-day plan and start again. By taking this approach, you should document how differently you would want to spend time and, importantly, where you were not planning to invest it.

My recommendation is that you call this new plan the **next 90-day plan**. You can use it as a self-control discipline to monitor your ongoing progress and keep any concerning issues foremost in your mind. Hopefully, this reset will allow all the *show stoppers* to be addressed or at least remediated.

As I will always advise, we all need to work, but it should be for fun!

Summary

The first 90 days are the foundation that you have to build for a successful start to your CIO career. These first three months will go in a flash, and if you don't plan this well, you will be diminishing your chances of success. It is easy to be distracted by events that occur and then require your attention. For your new CIO role, it is likely that the prior incumbent may have vacated the role already, and there is some extended clean-up required. Yes, this is frustrating, as you don't get credit for doing this, but it is necessary to set the tone of what you expect from your team.

In my CIO 90-day plans, there were other events that occurred that were not listed specifically in the plan. This will always happen, and you can't control everything, but you must try to keep to several key focus areas in your plan. Let me reinforce that your 90-day plan has to be detailed and it has to be shared. It is the start of your journey, and it will also allow you to reset the expectations of the team that you work with. In my view, it is impossible to be successful in any transformative role without getting the team behind you and aligned with what is required. Being aligned should never mean that they simply follow your instructions. As a leader, you want the team to themselves be leaders; the best results come from robust discussions in which a path of action is agreed on. Being open to input and getting everyone behind a decision is critical.

In the next chapter, I will explore the first 90-day plan for the new CISO. For some CIOs, they may also, in effect, be the CISO. Thus, this chapter is relevant to peruse.

6
CISO – The First 90 Days

I referenced in the previous chapter the backstory of how I prepared a plan for the first 90 days as groundwork for one of my CIO roles. This template has been reused several times in my career. As I approached my first CISO role, my plan was to logically reuse this "tried and tested" approach. Although, I also planned to make some tweaks, given that the expectations of each role have some subtle differences. I realized, however, that simple tweaks would not be sufficient.

A new CISO role comes with much anticipation. The role of the CISO is critical for the success of an organization and simply can't be vacant. In my case, as I entered the building, the previous incumbent was performing two roles, as CISO for the Asia Pacific region and global chief security architect.

The CISO role is one that you may have coveted, and it must be remembered that what brought you here may not be enough to make you a success in this new position. That is why some planning to consider how you spend your time in the first 90 days is critical. There will be new skills, knowledge, experience, and behaviors that you will need to succeed. This chapter is designed to help you accelerate this journey with some guidance that has worked in my career.

> **The fundamental question that is to be addressed in this chapter is as follows:**
> How do I write my plan for the first 90 days as a CISO?

This chapter will take a bird's-eye view to explore the essential dimensions that you need to focus on during your first 90 days. In this chapter, we'll be covering the following topics:

- A brief overview of my CISO plan for the first 90 days
- Exploring *People* in the 90-day plan
- Exploring *Process* in the 90-day plan
- Exploring *Technology* in the 90-day plan
- Building your plan for the first 90 days

A brief overview of my CISO plan for the first 90 days

In this section, I will explore the elements that make up a 90-day plan for a new CISO. My intent is to provide some examples so that you can prepare your own plan, either reusing the template or taking this as a starting point. The structure has some similarities to the CIO 90-day plan and several significant differences. That's not surprising though as these are very different roles. However, the plan is also to be shared with all stakeholders, and with staff, so that they are aware of where you are spending your time. As with the case of the CIO plan, the guidelines for the CISO plan remain similar:

- The plan must fit on one page

- It must be clear as to what you will focus on and what you won't focus on

- It is designed to be openly shared (so avoid any confidential references)

- It should be structured with time-based deliverables (this provides you with pressure to stay on track)

- People, Process, and Technology are plotted on the *y-axis* (refer to *Figure 6.1*)

- The first 90 days usually focus on process and technology more heavily than people

- The plan should be as closed-ended as possible, with measurable actions at the end of the 90 days

- Track your progress as the plan is completed (remember this is a branding exercise too)

- Finally, it is not supposed to be perfect – so revise it as necessary

Many of you will be visual learners, so I would recommend that you review the 90-day plan template as laid out in *Figure 6.1*.

The CISO 90-day plan will vary depending on the circumstances you are in. If there has been a recent breach, then this will be the focus of your time, reviewing the architecture of the layers of defense and how your controls operate.

Most new CISOs will want to combine several key assessment exercises to do the following:

- Understand the security posture

- Build their business understanding

- Assess their new team

- Review the IT and cyber strategy

- Understand regulatory requirements

- Understand the security and operations baseline

Figure 6.1 shows my 90-day plan, which has three *People* aspects and four for *Process* and *Technology*.

	30 Days	60 Days	90 Days
People	**What does success look like?** Ensure you have clarity of expectations from the CIO and the COO. Get support for this plan.	**Engage executive stakeholders** Share with the CIO and the COO your assessment of the situation.	**Contract success** Ensure specific support is secured.
People	**Understand key players** Do stakeholder analysis and influence map. Ask execs 3 things about their key issues for support and advice. What would you do differently?	**Start to build relationships at the exec level** Understand what their pain points are.	**Build coalitions at the exec level** Look for examples where 'no' is the appropriate answer.
People	**Assess the Cyber team** Evaluate the current team and culture.	**Build your future team** Identify talent gaps and take selective decisions to remediate the team.	**Build your team** Start to take action on decisions with your team.
Process	**Understand how Cyber engages the business** Explore engagement and governance gaps.	**Consider alternatives and options** Strike the right balance (evaluate where rigor and discipline is lacking).	**Take action** Implement deliberate efforts on governance.
Process	**Establish risk metrics** People respect what you inspect. Ensure this is well known.	**Monitor risk metrics** Ensure visibility of metrics and be committed to communicate actions.	**Refine metrics** Give recognition and market the results.
Process	**Accelerate business learning** Ask for access and inclusion to understand what external models your business partners value.	**Reflect on learnings** Be open to external environment and best practice benchmarks.	**Demonstrate learning** Share learnings with the business at meetings and also share these with your team.
Process	**Understand cyber governance** Learn how the management and the board operate.	**Review governance gaps** Share your proposed improvements.	**Execute improvements** Make immediate improvements.
Technology	**Review cyber strategy and strategic projects** Own this–explain intent, scope, and rationale. Evaluate the change portfolio.	**Challenge approach** Understand where changes may be required. Challenge what looks to be unnecessary.	**Get support from the CIO and the COO for any adjustments** Ensure airtime is secured.
Technology	**Understand security baseline** Be sure not to assume anything.	**Challenge security baseline** Be decisive.	**Take corrective action** Share this and repeat with the next issue.
Technology	**Understand security operations** Baseline your operations.	**Identify operational gaps** Measure and understand delta to best practice.	**Remediate and follow up** Show progress. Make improvements.
Technology	**Understand the regulatory book** Identify the compliance roadmap.	**Identify regulatory gaps** Understand whether there is over or under compliance.	**Adjust and follow up** Coordinate any adjustments with enterprise.

Figure 6.1 – CISO 90-day plan

You will note that this 90-day plan varies from the CIO 90-day plan, which has a marginally stronger *People* focus. This will always be a trade-off and it doesn't mean that *People* actions are less important to the CISO. The CISO will just have to work with *Process* and *Technology* more, in my view, to get to grips with the new role.

In the following sections, we will delve into each of the aspects of my 90-day plan. Let's dive into *People* first.

Exploring *People* in the 90-day plan

Traditionally, CISOs have not been renowned for their people skills. They are technology experts who understand complex architecture and cyber controls. Today's CISO is expected to be able to work well with the board, executive stakeholders, vendor partners, and staff, both in technology and business.

A new CISO is required to be strong in the *People* dimension. The role is no longer a back-office one but a strategic position that interfaces with external customers and regulators. There is also natural tension that CISOs and CIOs will face. Expecting this scenario, it's imperative for the 90-day plan to address it.

This section will also explore common-sense approaches to stakeholder management and how to anticipate how each party may react and respond to different scenarios. While this is not an exact science, homework will help you to navigate it more mindfully.

What does success look like?

Success will always be personally subjective, but what I mean here is how, as the new CISO, you can develop your *People* dimension so that it helps you to succeed. What success looks like is building strong, sustainable relationships with key stakeholders that will be strained in your role but never broken.

The CISO has many important stakeholders, of which the most critical stakeholder in the first 90 days is your boss. This will likely be the CIO, but in some cases, you may report to two people, including the COO.

In your very first face-to-face meeting, take the time to share your plan with your new boss and ask for feedback and input. The CIO will most likely have some potential quick wins that they want to be fixed immediately. Such quick wins, if possible, will be great for your reputation.

Tension with the CIO

Be aware that there will always be some in-built tension between the CIO and CISO roles as they will, at times, have different motivations. The CIO is concerned about "protecting" the enterprise, but they also are charged with enabling a change agenda.

One such agenda is vulnerability patching. The CIO will have a metric to have system availability at 99.999%. They will want to keep downtime to scheduled periods and avoid any potential disruption. Conversely, the CISO will want to expedite patching across the estate and there will usually be a lot of noise when they want to accelerate this activity.

The CISO will naturally be focused on cybersecurity and mitigating risks. While the CIO cares about these topics, they will want to prioritize digital innovation and, where possible, cost savings. These differences can lead to tensions between the CISO and CIO.

While this may not be obvious during the first 90 days, it is worth anticipating how you, as the new CISO, will navigate this minefield. The only saving grace is that you communicate and engage regularly with your CIO, thus, some level of collective decision-making can help keep both of you aligned.

The CIO will want you to engage with their executive peers and will undoubtedly have a long list of stakeholders that should be met. When you meet with these stakeholders, they will sometimes share and overshare items that are not in your remit. Be sure to pass the feedback on to the CIO. It is important that the CIO has your back, and you want to be a role model for this yourself.

Key action points

- Regularly engage with the CIO to foster open communication and establish a collaborative relationship to align priorities and decisions

- Recognize the differing motivations of the CIO and CISO and strive for a balance that considers both perspectives

- Pass relevant information to the CIO, even if it falls outside your direct responsibilities

- Remember that collective decision-making can help bridge gaps between the CISO and CIO roles

---

The relationship of the CISO with other stakeholders – COO and CEO

The obvious approach I advocate is to simply ask the stakeholders what their "hot buttons" are. Stakeholders include the COO and CEO, plus all the executive stakeholders.

Without some level of calibration of understanding, this can become onerous – your new CISO role will become the "messenger," and you will just be sharing information with different levels of the organization. While that is useful, it is probably not the best use of your time. Another area that can naturally be prone to tension is incidents. The CISO is usually informed of these events much more quickly than anyone else due to the operational nature of their role, whereby incidents trigger notifications from monitoring systems that the CIO won't always receive.

It is essential that ground rules are established from day one. When a serious incident occurs, the CISO must be clear about the preferred method of communication with the executive team. Is it a quick phone call or a secure message? Who has the responsibility of communicating with the COO and CEO? Once this is clear, then any potential misunderstanding can be averted.

Unfortunately, this sounds like common sense, but I have seen many good CISOs get grilled due to an incident and the sequence of communications. As cybersecurity continues to be a significant factor in reputation damage, it follows that having clear guidelines will keep you out of trouble internally and allow you to be able to focus on the actual incident.

> **Key action points**
>
> - Understand the "hot buttons" of the CIO, COO, CEO, and other executive stakeholders
> - Avoid becoming a mere messenger; instead, provide value-added insights
> - Define communication protocols from day one. Clarify whether quick phone calls or secure messages are preferred during serious incidents
> - Recognize that transparent communication contributes to safeguarding the organization's reputation

Understanding key players

The list of key stakeholders will be both broad and extensive. You will want to meet with your technology peers and key business leaders. There will also be risk, compliance, and audit peers that you should meet and establish good working relationships with.

I've seen examples where the CISO meets business stakeholders for the first time when there is a new incident and there may be limited contact and engagement. Clearly, this is not ideal. It is true that some CISOs may not be that active with their business engagement and this may be due to a strong focus on technical cybersecurity priorities over broader business discussions.

Indeed, the CISO will have to adjust their natural communication style to avoid using any tech jargon – this is not an excuse but a reality. In these business meetings, the new CISO should ask the same three questions:

- What are the key issues?
- What do they need support on?
- What do they suggest we do differently?

First understanding what the business priorities are and how the cyber team can support these initiatives will help the CISO develop their business partnering approach by understanding the security posture.

There will be times when the CISO will have to call on these relationships to drive the acceleration of a security program of work or a vulnerability patching exercise. Either way, the CISO will use this engagement to educate and brief their key stakeholders on why a particular incident may be important.

Once you have a "seat at the table," this can help by getting your team onto the business unit's planning table. This is when you start shifting from helping with "pain points" to beginning to build true coalitions, and the executive stakeholders will see this benefit as a "win-win."

I've worked in organizations that regard cybersecurity as a "lose-lose" situation. In those companies, the executives want to avoid being involved as they don't understand how it could benefit their business – instead, they try to minimize the impact by pushing back.

Key action points
- Identify and engage with a wide range of stakeholders, including technology peers, business leaders, risk, compliance, and audit professionals
- Tailor your communication to suit the style of your audience, avoiding technical jargon and focusing on business priorities
- Ask stakeholders about their challenges and priorities and inquire about their needs and how the cybersecurity team can assist
- Show how cybersecurity benefits the organization, turning it into a "win-win" situation

---

Using the Myers-Briggs framework to gauge how you can work with colleagues

We have committed to considering some alternatives and then taking action with your stakeholders. To do this, the CISO should complete some analysis and a formal stakeholder analysis and influence map. I won't repeat this as this was already detailed in the previous chapter. However, what I would suggest is that another simple way to understand your stakeholders is to plot them using the Myers-Briggs framework or any of the other commonly used personality prediction models.

Using the Myers-Briggs personality indicators, shown in *Figure 6.2*, is a common approach used to report psychological preferences in how people perceive the world and make decisions. The exercise roughly assesses your stakeholders – again, this should be a confidential assessment.

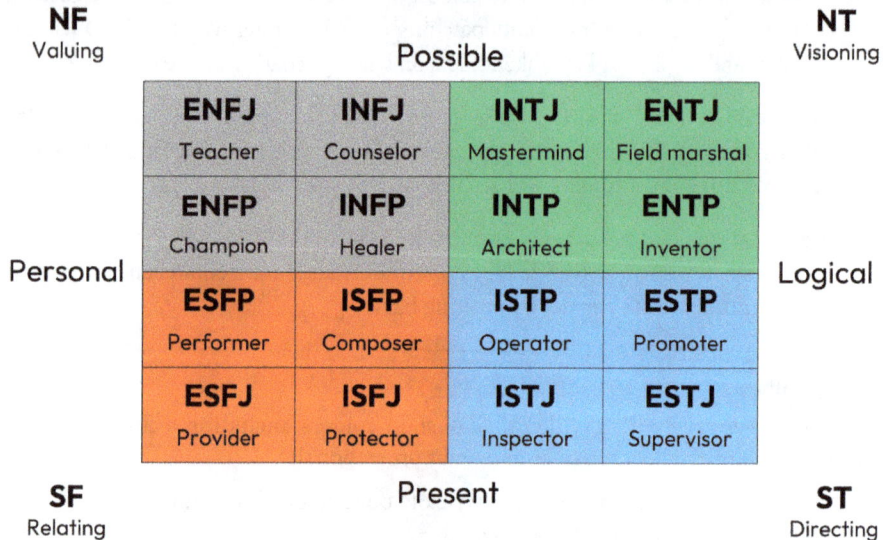

NF
Valuing

Possible

NT
Visioning

ENFJ Teacher	**INFJ** Counselor	**INTJ** Mastermind	**ENTJ** Field marshal
ENFP Champion	**INFP** Healer	**INTP** Architect	**ENTP** Inventor
ESFP Performer	**ISFP** Composer	**ISTP** Operator	**ESTP** Promoter
ESFJ Provider	**ISFJ** Protector	**ISTJ** Inspector	**ESTJ** Supervisor

Personal

Logical

SF
Relating

Present

ST
Directing

Figure 6.2 – Myers-Briggs model

A rough assessment can be done for most stakeholders by using a few simple questions:

- Are they extraverted or introverted?
- Are they intuitive or sensing?
- Are they thinking or feeling?
- Are they judging or perceiving?

This can be done in a closed workshop with the new cybersecurity team. It will provide some "Aha!" moments when you realize that these stakeholders may be being approached in the wrong way or perhaps by the wrong persons from your team.

Having used the Myers-Briggs approach to plot the key internal stakeholders, it is insightful to reflect on how they like to engage and who their natural peers are. This analysis should be extended to map how your lead CISO team communicates with these stakeholders.

It is very likely that your team will not have a direct mapping with the stakeholders and that is where you may need to engage the CIO and other parties to assist you.

In my own case, I had plotted some of the key stakeholders in my organization as follows:

Myers-Briggs indicator	Distribution of stakeholders	Stakeholder engagement tips
Extraverts	CRO, CEO, and CFO	Engage in lively conversation and provide opportunities for social interaction.
Introverts	CIO, COO, and CMO	Respect their need for quiet reflection and focus on one-on-one communication.
Sensors	CFO and COO	Present concrete evidence and practical benefits.
Intuitive	CIO, COO, and CMO	Discuss possibilities, share your vision, and spark their imagination.
Thinkers	CIO and COO	Use logic and objective arguments, and avoid emotional appeals.
Feelers	CRO and CFO	Consider their values and emotions and build a personal connection.
Judgers	CRO and CIO	Highlight structure, deadlines, and clear plans.
Perceivers	CFO, COO, CEO, and CMO	Be flexible, open to changes, and encourage their input.

Table 6.1 – Myers-Briggs engagement approach

This analysis, along with your own team doing the Myers-Briggs assessment, is a perfect activity for your first offsite meeting with your team. Your lead cybersecurity team will benefit from seeing how each of their peers operates – they could make some educated guesses, but it will not always be obvious.

Having your team also plot this chart (in a different color) can be an insightful exercise to see how relationships are mapped and where there may be some areas to shift to find greater synergy for individuals on your team. This will be a great stretching objective for your direct reports, and many may not have engaged with some of the key executive stakeholders extensively in the past. You may decide, for different reasons, to not follow the Myers-Briggs framework robotically and instead match your team based on your early assessment of how personalities will fit together.

As you will be working closely with your new team, there will be an expected and natural requirement to assess for yourself what talent you have inherited.

Assessing the cyber team

How you assess the cyber team is very much a personal approach. There will be opportunities for you to see individuals in action in one-to-one meetings, as well as in group sessions. Your observations on how they operate to solve problems and how they communicate will be areas that you will want to take special note of and be concerned about. You will likely also have reviewed their formal CV and received solicited and unsolicited feedback from stakeholders and colleagues.

There will be an overall sense that you gain in the first 90 days of their strengths and weaknesses, and some areas that they may need to work on. While you are doing this assessment in your head, the staff members will be conscious that this process is underway and they will want to understand your expectations.

At the end of the day, you will have an impression of each staff member – how capable they are and whether they are ambitious. Moreover, you will have a gut feeling about whether you can work with the person or not, and that will be based on whether you feel that you can trust them.

As the new CISO, there will be a "reset" that occurs with this transition and all cards will be on the table. Some staff will feel insecure as their past deeds and deliverables are unknown. The CISO must set their new team up for success and this assessment is the beginning of this exercise.

> **Key action points**
> - Observe team members in both one-on-one meetings and group sessions and document your observations to track strengths and areas for improvement
> - Assess their qualifications and experience based on their CVs and gather inputs from colleagues and stakeholders about team members' performance
> - Develop an overall sense of each team member's capabilities and limitations
> - Clearly communicate your standards and expectations to align everyone

Building your future team

Once this assessment is complete, you must share with your boss and then with your staff what your plans are. All parties will be awaiting this communication and will want to provide some input.

The new team may require some different resources in terms of skills, knowledge, experience, and behavior. This could necessitate some structural changes and perhaps some staff leaving their current role.

While this whole process can be unsettling and can impact morale, it is an essential part of the transition. As the new CISO, it is wise to be open to input but also to be firm when making decisions during this process. This will not be a democratic process and you, as the CISO, will have the say on any final changes. It is best to be crystal clear at the start that this will be your approach, and that you expect – once decisions are finalized – that the whole team will support any changes.

If new cyber roles are created and existing staff and external candidates can apply, this will be seen as both positive and negative by some. The new CISO can drive significant change if they are strongly supported by their new leadership team. But without this, the role will be hampered.

Patience will be required as, even once a new design is agreed upon and approved, the recruitment process is always longer and more protracted than desired and will extend beyond the 90-day period. In truth, if the new CISO can onboard a new direct report within 4 months of their start date, then that will be an achievement.

During this period of transition, the new CISO will feel hampered and will often be required to take on extra roles while the recruitment and onboarding progress. Thus, it can be a frustrating time, and the CISO should be mindful of this constraint.

> **Key action points**
> - Communicate your assessment findings and plans with your boss and the team and be open to feedback from stakeholders
> - Assess whether the team requires different skills, knowledge, or experience and consider moving staff to roles that better match their abilities
> - As the CISO, be mindful of additional responsibilities during this transition
> - Leverage your leadership to effect meaningful transformations

In this section, we have explored the *People* aspect of the 90-day plan, in terms of understanding how to build relationships with key stakeholders, the CIO, and colleagues. We have reviewed the Myers-Briggs model as a method to evaluate how your stakeholders naturally operate, and hence how you can best position your team with those parties. Then, we assessed and built out our new team. We'll now commence reviewing the *Process* aspect of the 90-day plan.

Exploring *Process* in the 90-day plan

By formal definition, a process is "a series of actions or steps taken in order to achieve a particular end." If you have just landed a new CISO role, you will need to understand how processes work – or perhaps don't work. The CISO will have a clear desire to have effective processes and be efficient, to avoid extra work and effort. For every new CISO, one of the most critical processes is business engagement. When this engagement is positive, the CISO can work with colleagues to understand business imperatives and how the cybersecurity risk posture sits in that overall context.

Understanding how Cyber engages the business

The CISO must play a strong leadership role to integrate the business strategy and objectives with cybersecurity. This requires the CISO to work with business colleagues to develop a cyber strategy that supports the business agenda.

Hence, ongoing dialogue is necessary to facilitate two-way learning. The CISO must ensure that the business understands that cybersecurity is everyone's responsibility and that the desired cyber culture requires all parties to take on this ownership. In addition, the CISO must learn the business language so that the value of cybersecurity in terms of risk reduction is understood and articulated in terms of the negative impact on the customer experience.

This engagement is ongoing and not just at the time of the development of a new cyber strategy. At any point in time, the CISO must have an awareness of where the business unit has the greatest cyber vulnerabilities and what steps are being taken to address this gap. They also need to be able to articulate how any new requirement from a regulator or that is identified by the board, management, or any other party, such as an auditor, will be prioritized vis-à-vis the current priorities.

There is one common approach used by CISOs to engage with the business, and that is to focus on the key risk metrics and progress against these measures. This is a data-driven approach that also should remove the technical jargon from the message. The new CISO can share these metrics with regularity to build credibility and trust. But note that there can be cases of metric fatigue, as the business can at times fail to understand why things have not already been resolved.

Key action points

- Work closely with business colleagues to align cybersecurity strategy with overall business objectives

- Create a comprehensive plan that supports the business agenda

- Ensure that everyone in the organization understands their role in cybersecurity and encourage a security-conscious mindset across all teams

- Use risk metrics to communicate progress and impact and regularly share relevant metrics to demonstrate effectiveness

---

Establishing risk metrics

Having strong governance processes, including cyber and resilience risk metrics, is fundamental to the role of the CISO. The CISO will also want to accelerate business learning to understand the requirements of the business. For every cyber team, there will also be frameworks adopted, such as NIST, MITRE, ISO, and so on. These frameworks will also provide a process for the new CISO to work with. Let's explore some of the key metrics.

For a new CISO coming into the role, there is nothing as important as establishing key risk metrics. But what is a good metric? That's a tough question, but here are a few key elements that you should try to put in place:

- Metrics must be easy to understand (avoid composite measures that confuse stakeholders)

- Metrics must measure risk and not effort

- Metrics should be forward-looking

- Metrics should be comparable (enabling them to be benchmarked with competitors)

The CISO is likely to have inherited a set of key risk metrics that they immediately don't agree are best practice. Thus, this must be a key focus at the beginning. For starters, many cybersecurity teams don't have data analytics as a core competency, and this may need to be built.

Patience is required as key risk metrics will take time – first, to agree on what is being measured and then to collect the data required. This will always take longer than desired and establishing new risk metrics in your first month will be a challenge.

As a starting point, just understanding what needs to be measured and what will not be measured is a great start. The CISO must look at a lot of details and then decide where there must be further analysis and development.

To start with, these metrics can be classified into smaller groups:

- **Asset management metrics**: Asset inventory is a foundation that is the heart of what the CISO must protect. This must therefore be accurate and complete, otherwise, the job will be more difficult:

  - Asset management inventory accuracy

  - Asset configuration management coverage

  - Percentage of assets not logged

  - The number of rogue (unauthorized) devices on the network

  - The number of asset exclusions (from standard protection)

- **Data security metrics**: Assets must be known and protected; this is again foundational:

  - Percentage of data assets not classified

  - Percentage of data assets not encrypted

  - Percentage of data assets unmanaged (with no owner)

- **Incident management metrics**: Incident management metrics can provide a measurement of how well the process works when things go well and don't go well:

  - **Mean time to detect (MTTD)** incidents

  - **Mean time to respond (MTTR)** to incidents

  - Number of data loss incidents

- **Vulnerability metrics**: Vulnerability management metrics provide a picture of your current risk posture, hence a "live" statement of what risk you are accepting:

  - The number of open critical vulnerabilities

  - The time taken to remediate critical vulnerabilities

  - Percentage of critical vulnerabilities patched within SLA

  - Percentage of code that was scanned prior to production

- **Patch management metrics**: Patch management metrics provide a measure of the trade-off between keeping production systems available and addressing known issues with fixes.

  - Patch coverage rate

  - Average time to deploy critical patches

  - Patch success rate

- **Third-party metric**: This measures the compliance level of third-party vendor assessments:

  - Third-party security assessments not completed

- **Privileged access metrics**: These metrics measure how well the principle of least privilege is being executed across the estate:

  - Number of inactive PAM accounts

  - Number of PAM sessions not monitored

  - Password rotation

  - Number of privileged escalations

- **Network security metrics**: These metrics provide a measure of how effectively the current network security controls are operating:

  - Number of firewall rule violations

  - Number of intrusion detection/prevention alerts

- **Risk culture metrics**: These metrics provide a view of the human vulnerability to click on links that appear to be genuine:

  - Phishing click-rates

  - Phishing remediation training rate

  - Phishing reporting rates

- **Threat intelligence metrics**: These metrics provide a sense of the coverage and degree of accuracy of the cyber intelligence process:

  - Percentage of relevant threats detected

  - False positive rate

- **Compliance metrics**: These metrics provide a measure of compliance with regulatory requirements:

  - Percentage of compliance with mandatory regulatory requirements

  - Number of overdue regulatory reports

- **Risk assessment metrics**: These metrics measure the day-to-day risk measures and how well the process is being performed:

  - Risk acceptance rate

  - Non-adherence to policies and standards

  - Overdue risk management issues

- **Cybersecurity investment metrics**: These metrics provide a sense of the state of completion of cyber investments and how much this is relative to the overall business investment:

  - The ratio of cyber investment to overall expense

  - The percentage of cyber platforms implemented

- **Cyber risk quantification**: These metrics use quantitative models to forecast potential cyber losses:

  - Annualized loss expectancy

  - Single loss expectancy (worst-case scenario)

Each of these metrics can be expanded on with further, underlying details. The trick is to understand what we have and what is missing to be able to grasp where the real risks are.

The CISO will have to reset the risk metrics, monitor them, and then refine them over time. There will be requests from the board and management for reporting and this must be "data-driven." This must be a priority and, as the new CISO, you must meet about this topic almost every week to ensure there is sufficient progress and focus.

These sets of metrics will be your security baseline and provide you with a sense of current practices and where the business is with regard to critical topics such as access control, privileged access, data protection and encryption, asset management, endpoint security, network security, incident response, and vulnerability management. You will want to be across all these metrics, and when this isn't the case, look to commission an effort to address it. Having these benchmarks is a great asset for a new CISO who wants to demonstrate their progress versus external competitors.

A new CISO will often get asked to compare the cyber maturity of their firm with other similar enterprises. Clearly, such benchmarks are useful, but not an exact science. There is always subjectivity that can be applied, hence I would caution against overreliance on such comparisons.

> **Key action points**
> - Create metrics that provide insights into future risks, which are easy for stakeholders to understand, avoiding complex composite measures
> - Evaluate inherited risk metrics and identify areas for improvement and consider building data analytics competency within the cybersecurity team
> - Use metrics to understand current security practices and gaps and meet weekly with stakeholders to track progress on projects and maintain focus
> - Use benchmarks as a reference, but balance them with context-specific factors

The next section explores how the CISO will need to accelerate their business learning to be seen as a credible leader. It is not enough to just understand cyber technology; there is an expectation that you understand the business.

Accelerating business learning

I have seen many examples of CISOs that don't care about understanding business processes. This has traditionally been due to the techy nature of IT security roles. However, this is no longer acceptable and there is a strong expectation for a CISO to be adept at technology and business.

In an age where cybersecurity is the number one threat to businesses, we should expect there to be open dialogue in both ways to work together to protect the enterprise. The CISO must build a deep understanding of business operations and not just the technology used.

The CISO must make sure there is a clear, strong alignment between cybersecurity and business objectives. This means that the most critical assets will be protected more than the less important ones. How they are protected in terms of cyber defenses should be correlated. The CISO must not only protect the crown jewels but also the "cash cows." These are the business groups that make the sales that support growth.

There will be cyber projects that positively impact security but perhaps drive negative short-term ramifications for business operations. The CISO and the business have to understand the risks and agree on joint decisions that, overall, make sense. As cyber threats are volatile and are always changing, this ongoing dialogue will enable new dangers to be addressed in a positive environment.

Taking the journey to learn about the business will be key to taking on the challenge and being able to reflect on what is needed. My advice is to invest time in learning what really matters strategically and operationally to your key stakeholders. This will require an investment of time to grasp the basics and a degree of curiosity to ask questions that may make you look silly. However, the impact on the business partners will be that they will sense that you are genuinely trying to understand and learn. This will pay dividends for you in how you are seen, but more importantly, in how you act in the future given this context.

The acid test for a CISO is that the feedback from the business is that the team, including the CISO, has an understanding of business processes and what cybersecurity risks are critical. It is not necessary for the CISO to become a business expert with depth of knowledge. Another way to think about this is that the business stakeholders want to see your degree of business knowledge demonstrated in your behavior and how you act considering this foundation.

A new CISO operates within a risk management governance framework that provides the structure with which they can manage cyber risk.

> **Key action points**
> - Move beyond technical aspects and actively learn about business processes
> - Ensure cybersecurity efforts align with overall business goals
> - Understand the strategic and operational priorities of key stakeholders and ask them relevant questions to deepen your understanding
> - Engage in ongoing dialogue to understand risks and trade-offs and agree on security projects that balance long-term benefits with short-term operational impact

Understanding cyber governance

The CISO will both love and hate cyber governance. This process provides the CISO with a regular headache and workload to fulfill the requirements of various risk committees and control effectiveness groups.

Without good cyber governance, the CISO is at risk of not being aligned with business objectives. There needs to be clear communication across different business parts on managing business risks holistically and assigning resources accordingly.

The new CISO will come to recognize that cyber governance is an effective way to help protect themselves from certain challenges. My own experience is that I naturally disliked governance and found it to be bureaucratic. However, in my career, I have learned that good governance can help you, as a CISO, to drive accountability and get better results as an outcome.

Using internal frameworks to manage cybersecurity

Cyber governance is subject to an operational risk framework. This is owned and managed by the operational risk management team, which is separate from the CISO and, if you like, creates the rules that are to be followed by the CISO. The framework itself includes cyber and technology policies and standards. This operational risk framework provides the foundation for the risk committee.

The risk committee is the governance body that provides oversight of the CIO and CISO regarding technology risk and cybersecurity.

Managing risk appetite is a critical outcome of good governance. Risk appetite provides some boundaries for what the business is willing to accept and not accept. Thereby, it directly drives the strategic direction and choices that are feasible.

One of the most critical governance outcomes is managing risk appetite. The CISO will be asked by the board to work on managing the defined appetite. This is a difficult endeavor. To have a clear and exact definition of what the appetite is and what is outside of the appetite is a fine art.

Let's take an example: any company would prefer not to lose sensitive customer records. However, in an example where we lose just one record, it may not be outside of the risk appetite. Many companies will set this appetite based on expectations around the reputational damage that data loss would cause.

Some companies set this tolerance to align with regulatory requirements. So, if the requirements suggest less than 5,000 records is okay – not great, but okay, then this may be set as the risk appetite. However, for a private investment bank, a single record of a multi-billionaire client will most likely exceed the tolerance.

To manage this appetite, the board will request that **Key Risk Indicators** (**KRIs**) are developed so that the CISO can track progress. These KRI metrics provide a measurement of whether critical cyber risks are within the appetite or outside of the threshold.

The framework will also require that cyber control plans are developed that are the basis for managing the path to improvement. This will require oversight through governance forums such as risk committees and control effectiveness groups.

Each control plan will have corresponding **Key Control Indicators** (**KCIs**) that measure the scope and coverage of the risk. For the CISO, governance also requires strong attention to control management. The selected cyber detective and preventative controls are expected to operate effectively. Given that threats are constantly changing, the reality is that some controls will need improvement or perhaps be ineffective.

Reporting processes themselves through various risk committees and board risk committees is also a key focus for a new CISO. To tell the story of the cyber risk profile, the following must be kept in mind:

- Where the company came from
- Where the company is now
- Where the company is heading

This summary is critical for the board to comprehend. This must be presented in business language, and note the following:

- Where threats have increased since the last report
- Where risks have decreased (progress against plans)

Similarly, any gaps discovered, such as weaknesses in the risk appetite statement or the KRIs will invoke governance to have these changed with the board's approval.

When cyber governance works, it can protect the CISO from random challenges. In the cybersecurity environment, it is critical to work to a plan and not be subject to new requirements unless they're prioritized to meet a critical new threat. I've seen how good governance acts as a "shield" for the CISO, so I recommend that you embrace it as a friend.

Good governance also extends to regulatory compliance and the requirements to meet various external regulations and industry standards. As a new CISO, it is important to understand what requirements in various jurisdictions you are subject to.

There will be times when you must trade off managing risks and what is required for regulatory compliance. The best guide I can give you is not to be confused and think that these are equally important. In truth, defending against a cyber-attack is a real risk; meeting a regulator's demand is important – it is best to keep it within that perspective, as there are many stakeholders who become confused about what should be prioritized.

Key action points

- Establish precise risk appetite boundaries that align with business objectives
- Effectively report the cyber risk story to the board by explaining the company's historical context, describing the current cyber risk profile, and outlining the strategic direction
- Work according to a well-defined plan rather than reacting to random challenges
- Understand regulatory requirements across different jurisdictions

---

Let's explore industry frameworks that you should consider.

Industry cybersecurity frameworks and their importance to a CISO

One thing to consider is that in reviewing governance, your team will demonstrate its adoption of standard frameworks. It is likely that multiple cybersecurity frameworks will be used, however, you will want to ensure that the team adopts one approach as the primary method.

For most mature enterprises, NIST is the preferred approach. COBIT is often used for smaller organizations that wish to build their governance practices. There are also use cases for using MITRE and CIS with different team members, but having too many frameworks is clearly not recommended.

There may be a case to use different frameworks within specific teams within cybersecurity, however, for external benchmarking purposes, NIST is the de facto approach, and this is what is usually shared with the board and management for their mutual understanding.

For reference, the commonly used frameworks are as follows:

Framework	Description
NIST Cybersecurity Framework (CSF)	The one most used by enterprises and there are multiple benchmarks against this framework. The NIST CSF is provided by the National Institute of Standards and Technology. Most boards have awareness of this framework and have been briefed on using the *Identify*, *Protect*, *Detect*, *Respond*, and *Recover* functions. This approach is risk-based and quite complex to implement, with approximately 1,000 controls.
MITRE ATT&CK Framework	This is commonly used by Cyber Intel and Operations teams and is based on how adversaries use **Tactics, Techniques, and Procedures (TTPs)**. It includes 149 techniques.
CIS Critical Security Controls	The **Centre for Internet Security (CIS)** Critical Security Controls is an action-orientated and prescriptive framework that is relatively easy to implement, with just 18 critical controls.
ISO27001	This is the international standard for information security management, and as such there is external certification across 114 controls.
COBIT	This model is developed by ISACA and contains 40 governance objectives. COBIT is relatively broader and more complex than the others to implement.

Table 6.2 – Comparison of cybersecurity frameworks

While various cybersecurity frameworks serve distinct purposes, NIST remains the common thread for external benchmarking and communication with stakeholders. By judiciously navigating multiple frameworks, organizations can achieve robust cybersecurity governance while maintaining clarity and focus.

> **Key action points**
> - Ensure your team adopts one primary cybersecurity framework and avoid framework overload
> - Consider the practicality of implementing each framework
> - Choose frameworks that align with organizational goals
> - Adapt frameworks as needed to maintain effectiveness

We have explored in this section how *Process* can be critical to the success of a new CISO. Processes may need to be established and reset if inadequate. Governance and metrics provide vital transparency to the CISO. Without a good grasp of business processes, the CISO will not succeed. Having good processes in place will help keep the CISO informed and on top of their role. This will again require the investment of your time and my advice is to invest.

In the next section, we explore *Technology* in the 90-day plan. This is often the sweet spot for a new CISO, and you may be tempted to take shortcuts, however, my advice is this: don't take that approach.

As the new CISO, getting your strategy right at the start will be the roadmap that you will set for the next few years. The *Technology* focus will also require you to have a benchmark for your starting point, and commissioning a Red team is a great way to get that baseline.

Exploring *Technology* in the 90-day plan

When an enterprise has technology that is too complex, then this serves to increase the attack surface and makes the company harder to defend. CISOs often come with a strong technological background, hence tech is the natural strength of the CISO.

It is also critical to success in the role. When cybersecurity tooling has been applied thoughtfully with automation, it can serve to protect an enterprise effectively. However, too many cybersecurity tools also have downsides, such as having high costs and creating inefficiency in terms of how the team defends against attacks. Having a clear cyber strategy is a key success factor for a new CISO.

Reviewing the cyber strategy and roadmap

I spent some time explaining key components of the IT strategy In the CIO's first 90 days. To avoid repeating this, allow me to add further commentary on some of the key differences prevalent in the CISO role. The cybersecurity strategy is owned by the CISO, and it is a key artifact to understand. The cyber roadmap is part of the overall IT roadmap.

This indicates how interrelated cybersecurity is with the overall strategy. This means the CISO must start by reviewing the IT strategy to grasp the current landscape. The entirety of the technology used across the enterprise is equal to the cyber threat surface. Every system, user, application, device, gadget, cloud environment, API, database, and so on, is subject to compromise and is therefore in the remit of the CISO.

The CISO will be held accountable for the enterprise's defense. Unfortunately, this may include shadow systems and APIs that have been unapproved formally by the CIO or CISO. While I agree this is flawed, the key is that it is in the best interest of the CISO to ensure that the inventory of assets is complete.

Within the IT strategy, there will be business objectives and projects that support it. Having clarity of these priorities will help the CISO achieve alignment. There will be some trade-offs to consider around new and old cyber risks that are inferred by the IT strategy. The cybersecurity strategy must be read in the context of the overall IT strategy and not considered in isolation.

> **Key action points**
> - Recognize that the cybersecurity strategy is owned by the CISO
> - Read the cybersecurity strategy in the context of the broader IT strategy
> - Consider how cybersecurity supports business objectives within the IT strategy
> - Evaluate trade-offs between new and existing cyber risks inferred by the strategy

Building your own Cyber strategy

As you recalibrate the cyber strategy, there will be some plans that don't make sense. For instance, a business unit may have applications that are critical assets but have no current plans for MFA protection. The new CISO should take a fresh perspective on this and attempt to negotiate that MFA be planned for the next scheduled update. This proactive engagement is crucial and will build trust with stakeholders. While it may lead to some tension in the short term, this is the right thing to do, and the challenge is a worthy exercise.

These challenges will culminate in a revised Cyber strategy and roadmap. This exercise will provide a baseline of the cyber position and how the execution of these plans uplifts and improves cybersecurity.

For the CISO, this cyber strategy will feature on board and risk committee agendas going forward, therefore it is a key artifact. The cyber strategy is a highly confidential document that could be used as a playbook to "hack" your enterprise, and as such, will need to be access controlled. A SWOT analysis will be included to outline where there are weaknesses and opportunities for improvement.

I have listed a few tips that will help you craft your own cyber strategy:

- Decide on the most important cyber controls to focus on
- Decide what existing cyber projects should be paused

- Seek out strategic choices that have a "trifecta" impact (reduce the most risk, have business value, and meet regulatory requirements)

- Involve your stakeholders during the review and in the outcome of the changes

- Get the board's endorsement of the new cyber strategy

A good cyber strategy is critical; however, it must be able to withstand an actual "live-fire" situation. The best way to test this is through a red team assessment.

Key action points

- Review existing plans critically, especially where critical assets lack MFA protection

- Use identified gaps and improvements to shape a revised cybersecurity strategy and roadmap

- Evaluate ongoing projects and decide which ones should be paused or adjusted and seek choices that balance risk reduction, business value, and regulatory compliance

- Validate the strategy through red team assessments to ensure real-world effectiveness

---

Baselining with a Red team

One of the best ways to gather this data is to authorize and conduct a **Red team exercise**. As the new CISO, it is in your best interests to have these gaps identified in the baseline. The Red team exercise provides an understanding of how your controls and teams operate under "live fire."

Usually, these exercises are conducted by specialist external partners. Once they are provided with a target, they then conduct specific intelligence-led collection for their reconnaissance stage. The Red team will use various **Tactics, Techniques, and Procedures** (**TTPs**) to emulate the actions of a threat actor. While the exercise is both resource- and time-bound, there will always be some gaps that are unveiled. And for a new CISO, this provides an action plan to address and remediate issues, which is perfect timing for you as a newbie.

The red team exercise will enable you to take full ownership of the role. You will gain a full and comprehensive perspective of your current position and be able to revise this to plan strategies to improve the protection of the enterprise. There is an added benefit, as the cyber strategy and roadmap will require discussions to ensure that there is alignment with the overall IT strategy. If there are resources and funding required, then these should be agreed, and any should be adjustments made to support the new CISO's agenda.

Key action points

- Use the Red team exercise to uncover vulnerabilities and weaknesses in your current cybersecurity posture

- Based on Red Team findings, create an action plan to address identified gaps

- Allocate resources (funding, personnel) to implement necessary improvements

- Leverage the Red team exercise to take full ownership of the cybersecurity role and revise strategies as needed to enhance enterprise protection

This Red team assessment will provide you with input for your security baseline.

Understanding the security baseline

The new CISO must establish a security baseline. There will have been a review of the security controls when they held workshops to review the cyber strategy. However, a follow-up will be required to dig into the details and gain a deeper understanding of the layers of controls and how these controls perform when tested by "live fire."

What is needed is a comprehensive and detailed walk through the cybersecurity control plans with each of the control owners to gain a deeper perspective of the plan and the gaps that exist. In essence, this security baseline is the detailed plans that have been prioritized from the newly approved cyber strategy.

There is significant work required to complete this exercise, and it is possible that the tasks will extend beyond the 90-day plan. However, as the new CISO, this is worthy of your attention to understand the current baseline, which is in effect the cyber risk posture that the enterprise is facing.

The CISO will want to know that remediation of the controls is risk-based, with a focus on the most critical risks first. I have seen enterprises where the control owners have chosen less critical assets to address as they were easier to negotiate and therefore showed progress. However, as the new CISO, you will care less about appearances and more about real risk reduction.

The **security baseline** will include all aspects of your cyber controls, including asset management, vulnerability testing, patching schedules, MFA, email filtering, encryption management, penetration testing, firewalls, security awareness training, and so on.

Your challenge will be to ensure that the sequence of the control uplift is planned in the appropriate order and that there are defined key control indicators that will provide a measure of coverage.

Any appropriate corrective actions should be prioritized and reapproved through the control governance that we explored earlier.

Key action points

- Engage with control owners to gain a deeper understanding of existing cybersecurity control plans

- Ensure the security baseline covers all relevant cyber controls, addressing asset management, vulnerability testing, patching schedules, MFA, email filtering, encryption, penetration testing, and firewalls

- Prioritize actions that genuinely reduce risk, rather than focusing solely on visible progress

- Regularly review and adjust the security baseline as needed and ensure alignment with board expectations and strategic goals

Now that you have learned about the importance of having a proper security baseline, let's move on to security operations.

Understanding security operations

In an age of constant, 24x7 threats from nation-states and ransomware attacks that have crippled large enterprises with advanced cyber capabilities, there are **Security Operations Centers (SOCs)** operating and separating your enterprise from business disruption.

A new CISO will want to understand how well their SOC is operating and what further investment is required. This is the security operations baseline.

Your Security Operations baseline

The front door to the defense of your enterprise is the **Security Operations** team. This team exists to protect the enterprise from cyber threats. The SOC team has to face unpredictable work pressure and deals with stealthy adversaries who specialize in deception and camouflage. There is constant monitoring against motivated threat actors.

As a result, the SOC team has a high risk of burnout, with long hours of shift work and alert fatigue from the volumes of false positives across their screens. A new CISO must understand the SOC and how well it operates, and they must be mindful to not increase the stress levels that are already in play. This is the **Security Operations baseline**.

The SOC needs asset management. This review exercise will have already been completed to understand whether the company has a good handle on its most critical assets. Note that cloud environments can pose extra challenges for inventory systems due to their dynamic nature. The SOC will need to have ready access to this list of assets. When there is a cybersecurity incident, the SOC team will need to refer to this "source of truth" to understand the asset and all the related configuration information. This data provides an understanding of the bill of materials of assets, and thus the relationship between assets that may be impacted.

Thereby, the SOC can comprehend the most exposed systems and hence assist in prioritizing the subsequent actions. There have been times when third-party partners have been compromised because the SOC has limited inventory information, and this can hinder their activities.

But what are some of the metrics to measure your company's SOC performance?

SOC performance metrics

It is likely that the SOC has some existing metrics. As the new CISO, you will want to evaluate whether these are fit for purpose and whether any aspects are not covered appropriately.

The CISO will aspire for their SOC to operate effectively and continuously improve. This is an effort that will require some iteration and further engagement with your team to stage these new metrics.

In the first 90 days, a new CISO will want to establish new performance metrics for the SOC, and these will include the following:

- Mean time to detect
- Mean time to respond
- Logging coverage
- Incident resolution time
- SOC workload distribution
- Alert triage time
- False positive rate

For each of these, a target can also be committed to, and this can be monitored. The new metrics can be compared to peers in the industry to help drive best practice. Another domain that the CISO will want to assess is how effectively the telemetry process operates.

> **Key action points**
> - Review existing SOC performance metrics and evaluate whether these metrics align with the SOC's goals and effectiveness
> - Recognize that limited inventory data can hinder SOC activities and ensure third-party partners' assets are considered in the inventory
> - Understand how tasks are distributed among SOC team members and create metrics to monitor the time taken to assess and prioritize alerts
> - Compare created metrics to industry peers for best practices and define realistic targets for each metric

SIEM assessment

The new CISO will also want to assess the operation of the **Security and Information Event Management** (**SIEM**) system as it provides transparency to the CISO. This is the centralized collection of security data (alerts and logs) from application systems, networks, and infrastructure servers. The CISO will want to be sure that the data aggregation is working and that SIEM alerts are generated for the SOC to investigate.

There will be logs provided for firewalls, endpoint detection tools, and various vulnerability management tools. The SIEM must correlate data across different events and seek patterns that imply a threat. Thus, there is a need for the SIEM to be regularly reviewed and to identify improvements. Some key activities will be to fine-tune detection rules, consider reviewing unused rules, and ensure these are disabled. Also, the CISO should consider machine learning approaches that could assist in automating these processes.

In testing security operations, the CISO will want to road-test how the current security policies and procedures operate, and this may require some simulation exercises to be run. The current crisis and incident response plan should be tested in a desktop exercise to ensure that the steps operate as expected and ownership is clear when the Security Operations team and others across Cyber must act.

Most enterprises want to operate within the principle of **Least Privilege** for access control. Unfortunately, across both legacy systems and the cloud, many will struggle with maintaining this practice. The baseline will provide an opportunity to measure this at a point in time, noting that this should be monitored regularly.

Finally, benchmarking against industry standards such as NIST should be an annual exercise. This can provide you, as the CISO, with a sense of where additional resources and investment are required. We should note that threats can change and impact the risk position, and cybersecurity remediation may not proceed as planned.

> **Key action points**
> - Review SIEM Operation by assessing data aggregation, alert generation, and correlation capabilities and optimize rules for accurate threat identification
> - Test existing security policies and procedures and run desktop exercises to validate crisis and incident response plans
> - Prioritize the principle of Least Privilege and address maintaining this practice across diverse environments
> - Recognize that threats evolve, impacting cybersecurity remediation plans

In the next section, we'll explore how a new CISO will have to focus on regulatory requirements.

Understanding the regulatory book of work

The new CISO will be subject to the current backlog of regulatory compliance work. All regulators are under strong pressure to take tough action, across all sectors and geographies.

At the time of writing, there were 156 countries that have cybersecurity legislation, and of those, the majority have data protection and privacy legislation. As such, the CISO is subject to multiple requirements across each jurisdiction. This makes the harmonization of regulatory requirements hard, if not impossible, to be fully compliant.

Let's take an example of incident reporting – 156 countries will have different definitions of what incidents and requirements to report.

The starting position is to have an inventory of regulatory requirements across global and regional agencies. Once this is documented, the requirements must be interpreted, as unfortunately, they will not always be clear and intuitive.

You must partner with regulatory compliance and legal departments to assist in these interpretation exercises. This will then result in an overall assessment of compliance. Some of the requirements may create an impetus to revise the enterprise's policies and standards. The overall assessment of compliance will also need to be escalated to the risk committee and may then provide ammunition for further resources to support these efforts.

Stakeholder management with regulators is a key criterion for success. This is a tricky role to play and requires some planning and careful execution. As the CISO, you must always be respectful of the regulator. They have a role to play in the ecosystem, and keeping them comfortable is your role.

How to manage regulators

Every CISO must pay attention to what regulators want. These regulatory meetings and catch-ups need to be taken seriously, with rehearsals and preparation to ensure readiness for each event. The CISO will usually be the key spokesperson for these sessions. These may not occur in the first 90 days; however, I would say that it is ideal for them to occur shortly after you join.

The CISO should prepare for the questions that are expected to be asked. For the regulatory meeting and follow-up, this must be project-managed with all appropriate management approvals. The CISO will set the tone for the meeting by being a respectful and active participant. There is often much that can be achieved outside of official meetings, and that is dependent on establishing a positive, trusting relationship.

My personal rule is to always be totally straight and honest when dealing with the regulator. You must be pleasant, respectful, and friendly, and when your enterprise is not compliant, be open about this. The regulator will want to understand that you have an action plan to address the gap. They understand their requirements are additive to the change agenda. Therefore, talking about broader mitigation to reduce risk with other work underway is relevant and will be comprehended.

We must always hold the perspective that risk management to protect the enterprise is the number one priority for the CISO. These regulatory compliance requirements are point-in-time obligations with a deadline set from outside of the company. In contrast, cyber risk comes from external threats, and these are constantly changing. There are deadlines set by internal management and they will always morph as threats change.

The CISO's career will be severely impacted by a cyber-attack. They are likely to be replaced following a serious breach or incident, regardless of whether it is the fault of the CISO. Similarly, regulatory non-compliance will be seen as a negative, however, the CISO may not lose their job over that immediately. Regulators have the power to enforce fines and penalties on an enterprise, rather than a specific individual. However, regulators have been known to recommend that the CISO and CIO be changed if they are not comfortable with them.

> **Key action points**
> - Document regulatory requirements across global and regional agencies and partner with compliance and legal departments to interpret requirements
> - Approach regulators with respect and understanding of their role and establish trust outside official meetings for productive interactions
> - Be honest and transparent when dealing with regulators and explain mitigation plans and timelines for addressing non-compliance
> - Recognize that risk management is the top priority for the CISO and understand that a cyber-attack can severely impact the CISO's career

We have explored in this section how a new CISO needs to focus on *Technology* aspects in their first 90 days. This will include reviewing and challenging the cyber strategy and roadmap and then having this new strategy endorsed by the board. They will then need to work through the security baseline and understand the state of controls and what will be uplifted. The new CISO must spend time with the Security Operations team and learn how the SIEM works and what metrics are measured. Finally, the CISO must grasp the regulatory agenda and what is required. How the CISO manages this relationship is vital for the enterprise and for the career of the CISO.

Building your plan for the first 90 days

As the new CISO, there is a real need to get the balance right across the abovementioned three dimensions – *People*, *Process*, and *Technology*. You won't succeed if you don't strive to fight your own natural instincts. We all are naturally more gifted in one of these dimensions and this will be your strength. While this will serve you well, this new position requires broader mastery.

The suggestion I make to you is that you develop your own 90-day plan after you have read this whole chapter. Once you have understood the end-to-end context, it will make sense to put "pen to paper" and start drafting your own plan. I discussed in the previous sections the intricacies of my 90-day plan template, which you can reuse or modify as you see fit.

My CISO 90-day plan was to both guide myself and inform my stakeholders where I planned to spend my time. By default, it also detailed where I wouldn't be spending my time, and that is also instructive to the broader team and stakeholders.

A 90-day plan helps you to set the tone and, therefore, is critical to your success. How you use your time is key and it will be necessary for your executive assistant to assist you in protecting your diary – to allow "think time," so that you are not overwhelmed with unnecessary meetings.

Summary

There is no honeymoon when you are the new CISO. It is incredibly important to get the first 90-day plan of the CISO right. There will be pressure to be on top of everything. It is therefore easy to get overwhelmed with activity.

Taking the time to work through their 90-day plan will help a CISO have a stronger probability of success in the role. While there is a much stronger focus on *Process* and *Technology* compared to the CIO 90-day plan, the CISO will still need to spend a considerable amount of time on *People* aspects.

The CISO will need to understand both the IT strategy and the cyber strategy and roadmap. These are strongly intertwined with priority and resource interdependencies. Also, there is a strong need to dive deeper into understanding technology and cyber controls. They also need to start building business relationships and there will be more time to invest in and build those relationships after the first 90 days.

There will be moments of truth where CISOs and CIOs find themselves in difficult scenarios. Those times can be incredibly testing, and therefore a time to accelerate your learning. We'll explore some of those scenarios in the next chapter.

Part 3:
Being the CIO or CISO

In this part, you learn about how you live up to the title of CIO or CISO. There are specific moments of truth that will occur in your tenure that will brand you within your teams and to your key stakeholders. How you show up as a *leader* will define your legacy and the brand that you have been aspiring to create. Some pressures come with the territory, and this includes how the CIO and CISO interact. To explore survival skills, we cover stakeholder management and how to build alliances.

This part has the following chapters:

- *Chapter 7, Moments of Truth (When You Accelerate Your Growth)*
- *Chapter 8, Understanding the Pressures CIOs and CISOs Face*
- *Chapter 9, CIO and CISO Survival Skills*

7

Moments of Truth (When You Accelerate Your Growth)

In every CIO and CISO role, there will be a moment of truth, where you can establish yourself within the role by your own actions. This will be a moment when you are truly tested and you may not feel ready and comfortable in your role, but you are faced with a challenge.

How you act and show your leadership will be a defining moment for you, and this will start to become part of your brand so that you can operate effectively under pressure. In essence, these moments are examples of using your behavior to provide leadership that makes a real difference. In these scenarios, there are some situations where you can accelerate your growth.

> **The fundamental question that is to be addressed in this chapter is as follows:**
> How do moments of truth accelerate my growth?

In this chapter, I have outlined a few examples of actions that I took in my first 90 days that helped me be successful in my new CIO and CISO roles. These actions will often occur during this time and could be extended as required into the next 90 days. The key is what you do and how you show up.

We will be looking at the following crucial moments:

- Building a team
- Building a partnership to deliver
- Handling a critical *hygiene* issue
- Dealing with aftershocks
- Having a sense of duty as opposed to loyalty
- Dealing with your first cyber attack
- Building a risk culture

- Being totally honest
- Getting the CISO and CIO aligned

Building a team

At CUA on my very first day, I learned that there were 90–100 resources assigned to the transformation program, which included contractors and permanent staff. When I asked, "How many of them are my staff?" I was told that only a handful had taken this on. It was very clear that there was much *spectator* behavior, and this project, which had just started, was doomed to fail.

I held a town hall meeting on that first day, and my IT team and the transformation team joined the session. In that introductory session, I did not pretend that this was going to be easy and that getting this completed successfully would challenge each of us. Indeed, I noted that the journey would require us to all get "comfortable being uncomfortable," and perhaps some should self-select and leave rather than endure it. I did also encourage the team, and I was committed to their development of new skills, and this was part of the transformation that we needed to achieve.

The town hall meeting was rather a shock, I think. Some were smiling and welcomed the tectonic shift, but many were uncertain and concerned. This represented the first step of my 90-day plan to assess my future team. In this case, I had to rebuild most of the team, as many contractors were involved, and the vacancies allowed me to find transformation-spirited leaders who wanted to join this cause and challenge. Being honest and transparent has become part of my personal brand. To me, this is the easiest approach, rather than avoiding difficult questions. I'm not sure how you can start to remember different answers, or should I say white lies. It is easier to be frank and honest. This can be uncomfortable; however, I feel that this is just part of the process required to get to the right outcome.

> **Key career tip**
> Get comfortable being uncomfortable.

The key lesson is that as a CIO or CISO, you cannot succeed without building a great team. A reset will be required for you to set the right set of expectations for the entire team as a group and for everyone in that group. When you are in your role, this could be after a few days or a few weeks – that is up to your own sense of how this change will impact others and the noise that it may generate.

For every new CIO or CISO role that you accept, there is a usual assumption that you can rebuild to a certain degree, but this is not a mandate for wholesale team restructuring. The amount of restructuring permitted is generally related to the amount of severe disruption to operations and projects it entails. Hence, your focus must be on resetting the team to your new requirements.

Remember that the danger is if you leave this too long, then it can increase the risk of your not being accepted. As a new CIO or CISO, your exact words will be remembered by the team very clearly in those first few encounters and meetings. It is indeed a moment of truth.

As the new CIO or CISO, you will always be leading a transformation, and getting your team behind this effort is always going to be a challenge.

Now that you have understood the importance of building the right team, let's look at another crucial moment — building effective partnerships.

Building a partnership to deliver

In those first few weeks at CUA, I also learned that even though we were in the early requirement phase, CUA had already sent three legal letters to **Tata Consultancy Services** (**TCS**) regarding their concerns about them not fulfilling their obligations.

This was the first time that CUA had worked with an offshore team. Indeed, it was their very first instance of working with an Indian vendor. I asked the CEO to allow me to reset the relationship. I noted that India is a very complex place, with class, religion, and caste differences that are remarkable. One example I quoted was that there were more than 1,000 castes in India and that society is built very differently, making it hard for Western managers to truly understand the complexities.

However, I noted that the legal letters had to stop immediately, and instead, we should have a respectful dialogue. While CUA was the customer, TCS was a giant in the tech world, and the shadow it cast was much larger than our own.

Perhaps there was an undertone of distortion, and any Indian contractor or vendor that worked with the organization was assumed to be from TCS. Perhaps this was not intentional, but it was clear and needed to be reset in order that the team could form, and the output algorithm maximized. Let's remember that this was the first experience that the enterprise had had with an overseas vendor.

What I established was a vendor partnership relationship that had a project meeting on Tuesday evenings to escalate critical build items with the Global TCS Development Leader. This was followed by a meeting on Thursday evening with the TCS Consulting CEO and CUA CEO. The approach allowed the Indian offshore team to work their magic, and miracles did occur with regularity once a respectful relationship was established.

My team embraced the diversity, and during the project organized cultural events that included Indian food, henna painting, and traditional music. The teamwork and interactions naturally improved and any sense of division between teams all but disappeared.

I'm particularly proud of team members such as Brett Barber, Karin Muller, Ross McGrath, and Clayton Price, who led and sponsored these cultural interventions. We had never spoken specifically about these gaps, but it was understood, and as a team, we realized that these invisible divisions would not help with the success of the transformation that we all strived for.

Figure 7.1 – Stronger cultural awareness drives the team culture

As I fast-forward to the "go live" evening, when we cut over to the new core banking system, it was impossible to tell who was contractor, staff, or vendor. Everyone was working on the mission to make this transformation successful. All worked around the clock and the adrenaline was running high – it was a true partnership.

However, building partnerships is not the only thing that may prove challenging. There will be many issues that will need your attention. Let's look at how one should handle them.

Handling a critical *hygiene* issue

As the new CIO or CISO, it is wonderful if you can help resolve a long-standing issue and get a quick win. Often, these are simple things that get attention, and your team is noticed for acting swiftly. In the 90-day plan, you will have assigned this as one area to focus on.

As you do the rounds of stakeholder meetings, they won't be shy about suggesting some areas to address. You should manage expectations and make it clear that you are interested in understanding their concerns, but also can't promise that will be the one item that gets prioritized.

I've seen an example of a large financial transformation underway and taking up all the bandwidth and resources for other projects. There were many stakeholders who felt that their needs were not being met and that they were crowded out by the larger strategic priority. A quick win was to acknowledge that digital innovation is on the agenda, and it is part of the overall portfolio of using an agile approach.

A critical *hygiene* issue that I wanted to provide a case study on was Metlife Japan. I had arrived in my new role as Senior Vice President, CIO, and Statutory Director. There had been a major transformation project that had just gone live the prior month to my start date.

All the discussions were about what was next in the strategy to execute. Then there was a major incident, and the new system just didn't work. Metlife Japan had embarked on this very large transformation program over four years. There had been a series of CIOs that had come and left during this period. I had accepted a new job and was the fourth CIO in four years. Did this concern me? No, not really. It was a new territory, but with gusto, I accepted the challenge with an immediate *hygiene* issue to resolve.

I chaired daily crisis meetings at the start and end of each day. As the system had gone live, there was much loss of face, which is a serious matter in Japan. The meetings were in a combination of Japanese and English language, along with tech talk.

While I had seen a demo of a new amazing solution a week earlier, I had frankly been underwhelmed with the functionality compared to the degree of investment made.

The root cause of the issue was the focus of the early meetings, and in troubleshooting what was happening and why. It became very clear that key end-to-end performance monitoring tools were missing, and some of the vital middleware that had been introduced lacked IT support.

I made it very clear that I was not looking to apportion blame, and my only concern was to get the system working effectively again. As I was the new CIO, it was a baptism of fire for the new team, but I did not want to add more heat because of the significant pressure that they were already under.

In the end, we fixed the issue, and the new tool was able to operate as designed. No blame was leveled at the existing staff and no further postmortems were commissioned. The focus was to look forward and ask the team to support me in the new challenges. There is also no credit that was gained from fixing someone else's poor execution; that is your job and is what should be expected.

Even though the issue was fixed, it must be noted that this won't result in a quick win. You might also have to deal with the aftermath.

Dealing with aftershocks

Japan suffers from earthquakes and exists on active tectonic plates. The island's long narrow shape amplifies the impact of earthquakes. It is also surrounded by shallow seas that enable the tsunamis produced by earthquakes to generate destructive power.

A story that came 12 months after these first 90 days illustrates the aftershocks that occur and are akin to the earthquake analogy.

As the CIO or CISO, your job is to set the strategic direction for the next three to four years. This is associated with a strategy and architecture roadmap. I was always taught that strategy is about making choices. During the process of creating this strategy, there is cost estimation, an assessment of the degree of change, and so on.

The analogy used was t-shirt sizes to anticipate the planned workloads from each of these new initiatives. The IT strategy and roadmap had several large t-shirt-size projects in the new budget forecasts. I did see some very large numbers and asked the *dumb* question, "Why is there such a large backlog of capital expenses?"

After some digging, it became clear that the prior CIOs had delayed obsolescence plans for the past three to four years. It was also the case that the large transformation program had diverted monies meant for this remediation, regardless of whether they had been told to do this, or perhaps just didn't have a strategy that made the costs apparent. As the CIO, your job is not to gain popularity. In this case, these costs were subsequently approved after further discussions with management.

The lesson I always believe to be true is that courage is required for honesty and integrity. Being the CIO that fixes these obsolescence gaps will not get you applause, but it is your job and to have to serve the best interests that protect the enterprise.

Doing the right thing is what the CIO and CISO always strive for. It is not a popularity contest and should never be seen that way. This is a moment of truth, and you always want to be on the side of the facts. Facts are friends to the CIO and CISO. They may not always protect you, but they must be what you use as your compass.

While facts are a crucial compass for CIOs and CISOs, navigating complex situations often requires going beyond the data. Enter the importance of having a sense of duty as opposed to loyalty.

Having a sense of duty as opposed to loyalty

I've seen many examples in my career of executives and managers who appear to be taking the option to protect their manager over the company. These can be extremely sensitive examples where there is personal loyalty to your manager.

In some cases, this sense of loyalty comes from past working relationships. They were perhaps brought over or promoted by this leader. These bonds aren't obvious to the new CIO or CISO, and they can sometimes only come from building some personal rapport with stakeholders.

I recall a tricky example with another executive:

Scenario	Dialogue
Executive	"How is the transition to your new role and do you have any concerns?"
David	"I'm concerned about the progress and outcome of the transformation."
Executive	"What can we do to address this?"
David	"[Name] is not strong enough to lead this transformation. We needed to make significant changes in leadership to be able to be confident."
Executive	"That person is my partner."
David	(embarrassed) "I'm sorry, I was not aware that was the case, but it doesn't change my view that this is not a good fit."

Table 7.1 – Dialogue to do what is right

This was the right thing to say, but clearly, I felt awkward at the time. It would have been easier if the executive had disclosed the relationship in advance. To close the loop on the story, I was asked this same question by my boss a few weeks later and told him of the overall dialogue and the fact that I had not changed my opinion.

It is not about being political but transparency for what is required.

Unfortunately, I've seen many other examples of a misguided sense of loyalty that managers can have to their executives. One example at another company was a senior manager who provided unwavering support to their executive.

When I questioned this, the response was that in the future this person would sponsor them to be a CISO. This was clearly not how selection and succession plans work, and in this case, when the executive did leave, that person was not considered or even thought of as a serious contender. These stories are here to highlight the danger that you have with your career if you make the wrong choices.

I can understand that when you really want to be a CIO or CISO and you admire someone as a role model, you can overlook their weaknesses. The key to becoming a good CIO or CISO is to take the best parts of what you have observed from different role models and become that type of leader.

These are moments of truth, and whenever there is a hard choice, then what is right for the enterprise in the long term should be the first option. The short-term or accelerated paths or those that are political are not going to be healthy in the medium and long term. Long-term thinking to do what is right versus expediting the situation is the only real choice for the CIO and CISO.

How you deal with a cyber-attack is a true test of your leadership, and this is never going to be an easy task. In the next section, we explore this moment of truth.

Dealing with your first cyber attack

The new CIO and CISO will have many moments that test themselves. As I have noted, these moments will build your confidence and help define your brand. A real test is to be able to remain calm when you encounter a cyber-attack. As you are new in your role, there will be many aspects that you do not yet understand.

For each CIO and CISO, the first serious incident that you face is always going to be a terrific test of your leadership and how you can drive a positive outcome from a negative situation. I recall in my first few weeks as a CISO that there was an attack against the SWIFT payment network. There had been some significant SWIFT attacks at the Bank of Bangladesh, and this was running laps in my head.

As a new CISO, I did not have a relationship with the SWIFT CISO, nor had I met the payments executives in my orientation. This was all to change quickly. A crisis incident meeting was called, and few details were clear and understood. There were many voices on the call and colleagues that were unfamiliar.

My orientation had not yet extended to reviewing the cyber controls on payment systems; thus, the architecture was a black box. I was aware that there were five to six inflight SWIFT projects that were underway and that some of these new cyber controls had been established.

I felt calm and collected, but a little unsettled as I was observing more than leading. In the crisis meetings, I tried to chip in with questions, but in truth, the incident process was running me more than I was running it.

The payments stakeholders were calm and considered. They had witnessed attacks such as this previously and were not overly concerned. While that was reassuring, I learned lessons from this very early chapter. I quickly learned how critical it was to keep the executive stakeholders informed, and the Asia Pacific COO was provided with regular updates during the day. Of course, how these updates are cascaded to the many stakeholders is always problematic in cyber.

There is a need-to-know basis, and that is not consistent across different incidents. As the CISO, you want your stakeholders to be informed and not panicky. I learned that it is a constant challenge to get these communications right and calibrated.

The incident also made us see some gaps that existed with the Security Operations Center's engagement in the incident, and how they were not engaged in the initial crisis call. That had to be addressed going forward. I also realized how very lonely the CISO role can be when there is a serious incident. There are not that many people who can help you, and information is limited. While you can fear the worst, it is best to stay calm and keep communicating what you know and don't know.

While clear communication is vital during emergencies, building a risk culture fosters a proactive environment where we can anticipate and mitigate threats before they escalate. The CIO and CISO have a key role to play in making this culture real, and one that can assist with their own mission to protect the enterprise.

Building a risk culture

A risk culture is not an obvious item for the CIO or CISO to consider as a priority. Here's the definition from the Institute of Risk Management: "Risk culture is a term describing the values, beliefs, knowledge, attitudes and understanding about risk shared by a group of people with a common purpose."

I've taken a new role as CISO at HSBC. The interview process was astonishing as I met with Global CIO of Commercial Banking Wendy Wang, who was double hatting as Asia Pacific CIO. She was new in her regional role and rebuilding her team. Asia Pacific was then 20 countries that accounted for 80% of the global revenues for HSBC, so this was a strategic position that she played.

Wendy explained that she had Regional CIO for Retail Banking, Chief Data Officer, and CISO roles all open. During the discussion, I explained to Wendy that I had managed cybersecurity as a CIO but had never been a CISO or been full-time in this capacity. However, I had written over 150 articles on cyber and really believed the connection between digital and cyber was a critical one.

Interestingly, my observation is that most CIOs prefer to avoid cybersecurity. They see only the downside that it can bring. For me, you can't do digital innovation without strong security, so these two go together. They are Yin and Yang and go together. This is why the CIO and CISO career paths are so intertwined.

Figure 7.2 – Digital experience and cybersecurity must be in harmony

In that first month in my CISO role, Wendy requested that I run a 4-hour cybersecurity training session for 65 Asia Pacific COO and Chief Admin Officers. It was clear that their engagement with cybersecurity was critical and HSBC at this point was "out of risk appetite" for cyber. There was a cyber transformation program named CSMIP that was spending US$1B over 4 years to uplift all the critical cyber controls across the group. However, the cyber team had not brought the business stakeholders along for the journey, hence they had concerns.

This unusual request served to put me into the position of being *comfortable being uncomfortable* as I had just 4 weeks to prepare. There were more questions than answers:

- How was I to prepare for 4 hours of content for such a difficult group of stakeholders?

- Did I understand all the HSBC controls?

- Did I understand cyber well enough in the first 90 days?

- What would be the hardest questions to address?

- Where did we need to show more progress in our cyber transformation?

My first instinct was correct, and this was to fully exercise the team and ask my new cyber staff to support me in doing this. I quickly learned that they did not have any experience or exposure to this level of management and were somewhat anxious.

We met to work out an agenda that was interactive and educational. I left the last hour to be Q&A, where I would stand in the lion's den and take questions from these COO stakeholders on cybersecurity.

The introduction set the scene. As the new CISO, I wanted to protect HSBC from being breached, but I could not do this without everyone helping this mission. I recall saying "We are in this together, and we will be stronger if we understand more and therefore can ask harder questions."

For myself, standing on stage with a microphone and getting questions without notice was not at all nerve-racking. I enjoy the performance and have the confidence to manage the process. This never means I must know all the answers. Instead, I must be confident but humble and take difficult questions. The audience understood I was new in my role, and this set the tone for how we would operate.

The cyber briefing session was a huge success, and I was asked to repeat this 4-hour session with the CIOs across the region and later with Regulator Compliance teams. Indeed, it became our template for providing briefings to Regulators and their teams when they requested in-depth sessions.

Most importantly, we used the momentum to establish this as a regular monthly cyber briefing that was attended by 150–180 executives for an hour. We had the COO, CIO, 2LOD, and other internal stakeholders attend and continue to ask questions. The same approach applied, and we had some slide materials, but most of all we wanted straight talk to share our concerns and where we needed assistance and focus.

For the cyber team, this represented a moment where they had Executive and Management exposure that previously they had never been exposed to. I had made it clear that I had their backs and would address any difficult questions and that their support was important to me.

As a new CISO, I was able to use this exercise to build individual relationships and trust with these staff members. The trust we built was so important because we would have to work on incidents and urgent regulatory remediation.

We had team members that covered cyber operations (Johnny Leung), third-party security (Simon Blanchet), the Hang Seng Bank CISO (David Chan), Regulatory Compliance (Madoc Tiong), the HSBC HK CISO (Albert Kwok), HSBC International (Teddy Leung), and HSBC India Hub (Pravesh Sharma), a great team that both challenged me and taught me. We made each other better. To me, that is the algorithm of what makes a great team. It is rarely a team of absolute star performers, but a team that is committed and willing to learn from each other but also has the teamwork to lift each other to work at the highest levels.

What we had accomplished was to build a strong cybersecurity culture where the broader team understood that we were in this together and protecting the enterprise was all our job. We had developed "an understanding of risk shared by a group of people with a common purpose."

Cybersecurity can't be assigned to the CISO or CIO alone; as leaders, we must engage the larger team to feel they have a stake in this shared agenda.

Empowering a broader team to contribute to cybersecurity requires fostering a culture of transparency, where being totally honest about threats and vulnerabilities is paramount.

Being totally honest

During my career, I have been well known to be straight, perhaps too honest. I recall a CISO scenario where I had joined a new company and was asked by senior executives to share my observations.

It became clear that they didn't want to hear negative observations. Of course, I had assumed that the senior executives that I spoke to were bona fide in their request. In different forums, I tested this and found their contradictions to be the culture that they had created.

A culture of blind loyalty and not being data-driven can be dangerous. As a CIO and CISO, you will come across these examples unfortunately too frequently. How you manage them defines your own leadership impact.

I find honesty to be an easy way forward, as the only real alternative is the opposite – to be dishonest. In truth, there are a few shades in between, but this still includes an element of deceit. As a CIO or CISO, it is just easier to be honest and direct. It otherwise becomes too hard to remember what story you said to whom.

Being totally honest is the approach that I have used to help get the CIO and CISO aligned. There are many reasons why they are not in alignment and can have objectives that appear to be conflicting.

Getting the CISO and CIO aligned

I've seen poor alignment between the CISO and CIO many times. We want these two executives to be working closely and effectively for the enterprise to function like clockwork. Digital transformation can often lead to natural tension.

The CIO is responsible for delivering the digital customer experience. There is pressure to remove friction from the process and allow new software releases to occur as quickly as possible. In contrast, the CISO is concerned with ensuring that these new releases are security tested for any new vulnerabilities and that our threat model is not adversely impacted by these planned changes.

In the CIO's team, there are developers who are operating and encouraged to be creative in order to be innovative. Unfortunately, I've seen developers who have taken this definition too literally, and downloading the latest malware to learn how it works is just not acceptable.

I have also seen strong disagreements when it comes to APIs and **multifactor authentication** (**MFA**). As a CIO, there is pressure to provide business functionality to allow collaboration with customers, partners, and third parties. This pressure often requires a new API to allow information to be shared externally.

Taking the time to harden these APIs will be a challenge because the CIO will want to do this after it is released. Similarly, MFA on digital channels is where I have seen CIOs not being the strongest supporters, and this can lead to misalignment with the CISO.

Summary

There are moments of truth that define the CIO and CISO. This is indeed where they earn the big bucks and justify the value that we place on these roles. These moments are teachable points that can accelerate your own growth.

In each moment of truth, this is when the skills, knowledge, experience, and behavior can be stretched into uncomfortable positions. You will grow from each moment of truth, and it will become part of your experience.

The scenarios listed are just some of the potential examples that you may come across. There are some warnings that being totally honest will always be the best approach. Unfortunately, there are some who may confuse loyalty and sense of duty.

You must aspire to be an absolute role model, and that is not always easy.

Finally, I explored how the CIO and CISO can at times operate in an uncoordinated fashion, and their efforts may lack synergy. The CIO and CISO are natural allies to progress their agenda, but there are times when they may not be aligned.

There are key moments of truth that require you to be *comfortable being uncomfortable*. These are some of the key takeaways from this chapter:

- Building your new team is a key moment

- Team culture starts with respect for individual cultural differences

- Fixing a critical *hygiene* issue doesn't always mean a quick win for you

- Honesty and courage are required for the CIO and CISO role

- Be very calm in a serious crisis such as a cyber attack

- Develop a risk culture that engages your team and the broader team

We have seen that the roles of the CIO and CISO have moments of truth. Over the longer term, we do want the CIO and CISO to be able to thrive and not just survive these moments. These moments of truth can be stressful and make the roles of the CIO and CISO not the most fun. The next chapter talks about how you can better manage your own stress in these roles.

8
Understanding the Pressures CIOs and CISOs Face

It goes without saying that the roles of the CIO and CISO are not easy. These are high-pressure jobs that are handsomely rewarded. There have been significant shifts in the requirements for the role over the years.

The reality is that the job description for these roles is not a true reflection of the role. These job description artifacts are built and required by HR but are not referenced once you are in the role and operating.

When you are the CIO, there is an expectation that anything that has a power cord is a kind of technology and that you are dedicated to supporting it. I've had many colleagues provide feedback to me about audio-visual systems not working. The CIO also must take some of the "heat" for business-developed applications that have been developed outside of the IT portfolio.

Similarly, the CISO must manage their role carefully and not be expected to protect every device that has a power connection. For instance, the CISO is often called into discussions about fraudulent site takedowns and other fraud efforts.

My personal view is that a blank job description is the best. There should be no restrictions on what you can and cannot do. For some, that may be daunting, but as a CIO or CISO, there are fewer reasons to place constraints on yourself. Some of you will be intimidated by an empty job description. For me, it is liberating; I would not like a job description that is too prescriptive as it would feel very restrictive.

In this chapter, you will explore the pressures that being a leader brings. There is a natural weight and burden that you carry in the role of a CIO or CISO. The average day in the life of a CIO or CISO is filled with activity and challenges. Their day is spent juggling these priorities, and that naturally leads to stress.

For each aspiring CIO and CISO, this stress is an aspect of the role to be coped with. This chapter provides insights into how this stress manifests and what you can do to cope with it and thrive.

> The fundamental question that is to be addressed in this chapter is as follows:
>
> How do I manage the stress that comes with the CIO and CISO roles?

The stress that you feel can depend on the type of CIO or CISO that you are. There are different types of CIO and CISO – each comes with different qualities and attributes. For simplicity, I have added some persona names for the different CIO and CISO types, and they will each entail specific stress points.

The typical enterprise will have a CIO and a CISO, who then have to work with each other. They may be alike or they may be different – their interplay, or lack thereof, can create tension and stress for both parties.

In this chapter, we will be looking at the following:

- The weight of being a leader
- Exploring a day in the life of a CIO
- Exploring a day in the life of a CISO
- How the CISO and CIO can manage stress

The weight of being a leader

The CIO and CISO must make difficult decisions every day. Their average day involves hot topics with their teams, business stakeholders, management, and the board.

Most CISOs must confront significant business changes and have management pressure on them to meet their expectations. There is a large global cyber talent shortage, which will make the role even harder.

Given the natural stresses of cyber-attacks and defending the enterprise, the CIO and CISO will often feel unsupported and worry about their own personal liability should the company be breached.

It is important to note that stress itself is not necessarily a bad thing. A certain amount of stress can help CIOs and CISOs to perform at their best. On the other hand, when there is too much stress, this can lead to team burnout. The CIO and CISO are at the top of the food chain but have a disproportionate amount of stress on their shoulders.

Figure 8.1 – Carrying the weight of the world

The CIO and CISO will feel the weight of the world on their shoulders, as they must manage a complex portfolio of projects that have strategic value to the enterprise, while also keeping the business enabled with an excellent digital experience for internal and external customers.

Plus, when things go wrong, there are only a few colleagues with whom they can speak, as the issues and incidents are usually highly confidential and sensitive. Indeed, systems operations might have a material impact that involves stock market and/or reputational damage.

The weight of the job is something that you won't be able to fully comprehend until you are in the CIO or CISO role and carrying this load each day. This weight comes with the role; it is not that you seek this burden. To be a CIO or CISO over the medium-to-long term, this is a factor you must get used to and not feel weighed down by.

Not surprisingly, the average CISO tenure is two years, and for the CIO it is four years. These two facts alone provide strong evidence that managing your own stress and that of your team is a key objective for the CIO or CISO. In the average day in the life of the CIO or CISO, they have to deal with a multitude of issues. The next section explores an average day in the life of a CIO.

Exploring a day in the life of a CIO

The life of the CIO is both dynamic and challenging. In this section, we will explore the diverse nature of the role and the highly demanding requirements that the CIO faces in their average day.

The CIO is never going to be bored in their role; they have many demanding internal stakeholders who want to accelerate digital change, and they will never not have a backlog of additional projects. The CIO has a broader role than that of the CISO, so the CIO will be spread even more thinly across more domains and will usually have more business meetings.

A day in the life of a CIO

An average day in the life of a CIO is incredibly stressful; it is a different stress to that of the CISO, and they naturally spend more time with business stakeholders in governance settings. Yes, they also must deal with cyber-attacks, but they are one step removed from them, which can either mitigate or intensify the pressure that is felt.

In this section, we explore where the CIO spends their time on an average day. I would argue that the CIO, in contrast to the CISO, spends more time on strategic matters than at the operational level. However, they can get dragged into **business as usual** (**BAU**) when there are serious outages that impact production systems.

The CIO has a broader remit across technology, data, and cyber. Hence, they are stretched thin from both an intellectual standpoint and a time standpoint. While there are also different CIO persona types that may be in the role, the content of work should not differ much between them.

Regardless of their own personal preference, there is a strong expectation for the CIO to drive digital transformation. Thus, delivering strategic programs and managing resources across the project portfolio is a necessity. The stakeholders will want the CIO to help them drive transformation and innovation while remaining budget-conscious and expecting unit cost reductions. Hence, there will be architecture and cloud planning required to help deliver these outcomes.

A day in the life of a CIO

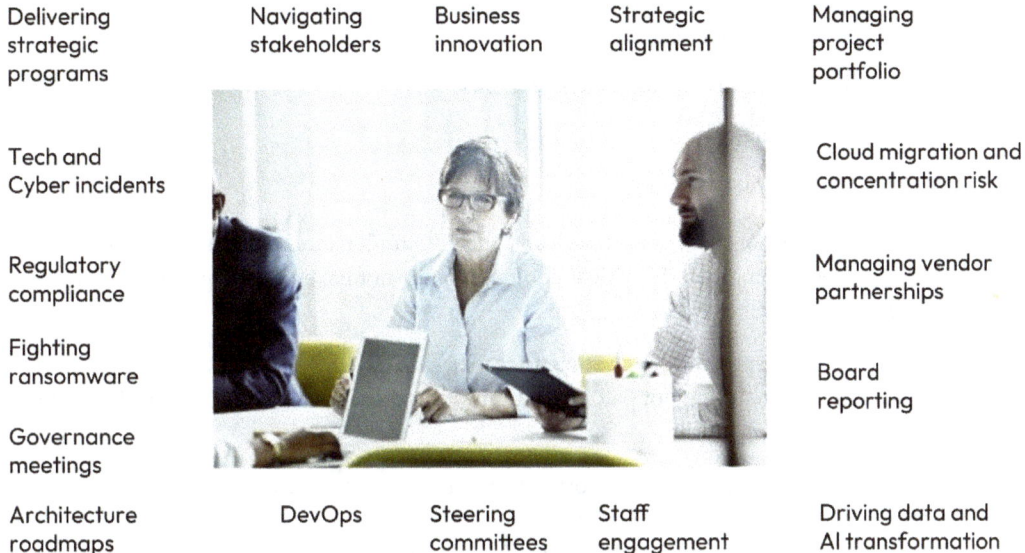

| Delivering strategic programs | Navigating stakeholders | Business innovation | Strategic alignment | Managing project portfolio |

Tech and Cyber incidents

Cloud migration and concentration risk

Regulatory compliance

Managing vendor partnerships

Fighting ransomware

Board reporting

Governance meetings

| Architecture roadmaps | DevOps | Steering committees | Staff engagement | Driving data and AI transformation |

Figure 8.2 — A day in the life of a CIO

The CIO will have stress coming from having accountability to drive innovation and help with the broader business transformation. This can mean that a CIO will feel this accountability to move fast but also to take care of cyber risks. A very good example of this is the introduction of **Multi-Factor Authentication (MFA)**; typically, this is required for externally facing systems. MFA creates a degree of friction in the customer experience, but this makes transactions more secure. The CIO will feel pressure, however, to not hinder business growth with such friction.

I do recall as a CISO having to *push* a CIO to adopt MFA more quickly, as we had seen some bot attacks. These had been targeting mobile banking solutions at a major organization that I worked for. The push-back not to do this across digital teams from the CIO was significant. It was clear that they were being rewarded for things other than cybersecurity.

In this instance, the CIO had somewhat conflicting priorities. The CIO wants to not disrupt the business by creating more friction, and they also want to improve cybersecurity – but they probably would rank those two objectives in that order.

These situations are particularly stressful when the CIO and CISO are not in sync. In many cases, the CISO will report to the CIO and that will only exacerbate this stress and pressure.

There are many other examples of when this occurs:

- Patching vulnerabilities

- Development pipeline code scanning

- Production data stored in test cloud environments

- Data at rest that is unencrypted

- Third-party partners' data retention

Technology is the lifeblood of most companies, and the CIO understands that their actions could simply stop the business from operating. Most companies may not have manual processes that can be reverted to in the event of business disruption.

The CIO is aware of this responsibility and has to walk a fine line between minimizing business disruption and being cognizant that not patching, scanning, and so on can create a cyber risk that is not within their risk appetite.

Having to operate in such mutually contradictory circumstances stresses the CIO, and this stress must be carefully managed. There are unfortunately few people that you as the CIO can share your concerns with, and usually, you will have to reflect on this solo. Yes, it is lonely, but it's part and parcel of your job.

CIOs face significant stress regularly. We'll explore the specific challenges that contribute to this stress in the next section.

The stress felt by different CIOs

The stress and pressures you face as a CIO will depend on your approach to your role. In the case of the CIO, there are six distinct persona styles, and each will come with a different stress. The question you may be asking is, *What kind of CIO will I be?* Your own natural approach to your work will give you this answer. It's best to find out that answer with the help of a mentor or close ally.

Depending on the type of CIO you are, you will face different types of stress. We will be exploring that in the following table:

CIO styles	Approach	Stress indicators
Transformational CIO ← →	• This kind of CIO relishes leading transformational change in their enterprise. • They are natural **change agents** who inspire themselves and their teams to drive large-scale IT programs that enable significant business change. • A strong vision and courage are mandatory.	• This CIO is a person who creates significant strain in their enterprise by leading disruptive change. Being a change agent can mean you create stress, and this can be redirected back to you as stress for yourself. • They will be driving toward hard deadlines and will often be unable to directly control many of the departments impacted by their changes. • They will be confronted with enterprise-level change resistance from staff, other executives, and potentially external customers.
Strategic CIO	• This CIO has an extra focus on strategic IT alignment with the overall business strategy. • Strategic CIOs are often instrumental in driving the company's digital transformation strategy. • Out-of-the-box thinking and strong stakeholder management are mandatory for the strategic CIO.	• Maintaining business stakeholder engagement and aligning IT with the business strategy is an endless job for many such CIOs. • In dynamic industries with regulatory changes and market shifts, this maintenance and alignment can be a formidable task. • This kind of CIO may experience more stress due to decision-making and risk management, as uncertainty is quite normal.

CIO styles	Approach	Stress indicators
Operational CIO	• The operational CIO is focused on ensuring that IT systems are running smoothly and efficiently. They are typically experts in IT infrastructure and operations. • Operational CIOs are often responsible for managing IT budgets and resources and ensuring that IT services meet the needs of the business. • Strong process thinking is mandatory for the operational CIO.	• This CIO runs the BAU of IT systems to ensure that these technology components run efficiently. • It would be fair to expect that their stress levels would be lower than other styles of CIO. They do, however, have higher stress levels, when asked to switch to a different CIO mode.
Business CIO	• The business CIO is often a former business peer who has been promoted internally. • The business CIO has often been placed into these roles due to the poor alignment that technology has experienced, leading to strategic misalignment. • This CIO cares about stakeholder engagement and how IT can strategically support the business.	• This CIO comes with established business relationships and an understanding of processes. • They may have less technology, cyber, and data knowledge, and this can cause some degree of stress. • There may be some temptation to engage external consultants to provide advice and counsel.

CIO styles	Approach	Stress indicators
Partnering CIO	• The partnering CIO spends their days meeting with business stakeholders, external partners, and their staff. • They want to build a trusted environment that provides a platform to enable business delivery. • They should have strong interpersonal skills and be organization-savvy to navigate all potential barriers.	• This CIO works with stakeholders and must listen to their requirements while balancing their own needs with them. • This approach attracts reasonable levels of stress as they must balance the needs of the stakeholders with the needs of their own team. • Their greatest stress comes when they have conflicting requirements and must resolve them.
Empowering CIO	• Empowering CIOs always look for the unconventional path. They can get their teams to work well together to bring forward ideas that may radically change the business process. • This style of CIO must be nurture-oriented, with an emphasis on lifelong learning to enable their team to continue improving and developing.	• This CIO creates a safe culture that allows their teams to fail. This approach may be viewed differently by stakeholders who might admire innovative thinking but be concerned about practical delivery. • This persona has arguably the lowest levels of stress.

Table 8.1 – CIO stresses by persona

Let me add that the style of CIO that you adopt is not always a conscious choice. It's determined by your natural approach and how you operate as a leader. What is important is that you understand what kind of stress comes with the territory and then take action to address it. Indeed, it would create more stress to try to act in a way that is very different from your natural style, so I would recommend against stretching yourself too much here.

The CISO and the CIO both feel the weight of their roles. This makes the work-life balance hard to maintain, and it must be actively managed to be sustainable. It is easy to perform heroics but I'm not sure that this is ever rewarded, to be honest. The only answer lies in managing your stress effectively. In a later section, we will be learning a few ways to do that.

> **Key career tip**
>
> Recognize when you are under pressure and make sure that your behavior has not changed. Being calm under stress has to be practiced.

Exploring a day in the life of a CISO

Just like the CIO, the life of a CISO is also dynamic and challenging. In this section, we will explore the diverse nature of the role and the incredible juggling of hot topics that's required during the average day. There is rarely going to be a boring, routine day where you will feel that you have complete closure.

How you as the CISO manage yourself is important to your success. The CISO should be calm in a crisis and think logically but also be very action-orientated when bad actors are attacking the company. Let's see what a typical day looks like for a CISO.

A day in the life of a CISO

There is no such thing as an average day in the life of a CISO. Each day will bring fresh challenges and a potential zero-day vulnerability. The role of the CISO is naturally stressful, as it entails working with a team that has a full remediation change agenda as well as working on external threats that keep increasing at an exponential rate.

The CISO will be constantly on the alert and learning about progress against external threats as well as having to deal with actual incidents (and false positives). There will be planned and unplanned testing of the team occurring with red teams and crisis simulations.

This is all against the backdrop of a global shortage of cyber talent. The CISO must accept vacancies or hire staff with less experience, skills, and knowledge than they would prefer. At the same time, the CISO must be extremely conscious of the stress levels of their own team and should not make them any worse than they need to be. Managing staff to be productive and reducing their burnout is your dual objective – that is not easy.

Boards and management are coming under more pressure to uplift their cyber maturity, with regulators globally enforcing more prescriptive requirements. This can add more work to your backlog and will typically override what is already in your backlog.

There are other times that the CISO will want to take time to plan out their roadmap and cyber strategy, or to develop a cyber risk quantification model. These require some focused time and more detailed analysis, which will be challenging if there are operational cyber incidents or urgent remediation that is also underway.

At another moment of the day, there might be a detailed discussion around cloud configuration and potential data security issues stemming from poor data classification. The list is endless and can be exhausting.

The CISO will have to learn to regulate their time to focus on what activities are going to drive risk reduction more than other tasks. Again, this must remain top of mind as the day unfolds – everything will be framed as time-critical and the topmost priority, but they can't *all* be so.

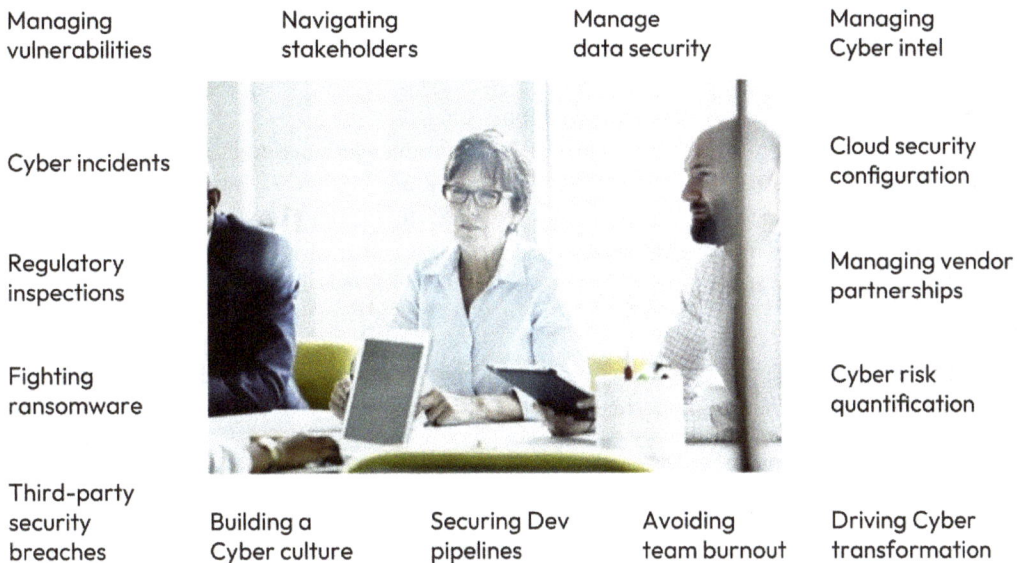

Figure 8.3 – A day in the life of a CISO

On an average day, a CISO will have to juggle many of the hot topics shown in *Figure 8.3*. Each requires management attention, and it is likely that some will have to be delegated. Yes, I do propose having a good delegation approach because it will help you grow your own team. However, if there are critical items, then you will still retain the stress of knowing that those items are with your team and not yet resolved. Some days will see the CISO spend all day focusing on just one hot topic. For example, a board or risk committee report requires significant bandwidth from the CISO.

I do recall moments when I was working on a board paper while there was also a regulator report being drafted, and we then had a suspected cyber incident. As a CISO, protecting the company is your primary role and you should most likely drop all other work and lead the incident response and associated communications with management. There will be other stakeholders who don't have to manage incidents, and to them, a board paper or regulator report is their number 1 priority. This can create extreme stress for the CISO, and this must be managed by educating others rather than not operating in the right order of priority. I can't imagine any more stressful moments for a CISO. You will feel pulled in three directions and know that only so much can be delegated to your team. Yes, this is a stressful job, and that stress needs to be managed by yourself.

A new cyber incident will deserve your full attention, and a board paper and regulator requirement, which won't disappear, can be delayed. Unfortunately, the sheer variety of challenges and everything being a high priority requires strong time management and personal resilience. As you can imagine, there is a lot of stress that must be dealt with regularly. In the next section, we will look at the various challenges that cause stress among different CISOs.

Stress felt by different CISOs

The CISO, I would argue, is a role that comes with more stress than that of the CIO. Having been both a CIO and CISO in my own career, this judgment comes from experience and seeing this play out. There are some CIO roles that are incredibly stressful and some CISO roles that are inherently less challenging. This may often be a factor in the industry that you operate in. Cyber threat actors have recently been more focused on hospitals, critical infrastructure, and governments, but this pivots over time, and stress can shift with these trends.

When you are the CISO, there are calls that can occur 24/7, including during holidays. Cybercriminals will often be active at 16:00 on Fridays; there have also been famous cyber hacks that have occurred on public holidays, as this will disguise their activities.

The CISO has to be on constant alert for what feels like attacks from inside (stakeholders), external attacks (cyber threats), and regulatory requirements.

Similarly, the stress and pressure that is felt by the CISO are also directly influenced by the style that they adopt. The CISO comes in roughly six different personas, and each approach has a distinctive stress that comes from that style. As for CIOs, the kind of CISO you will be depends on your natural approach, which you can find out more about from your mentor or close ally. We will be exploring the stresses faced by CISOs based on their persona in the following table:

CISO Styles	Approach	Stress Indicators
Technical CISO	• The technical CISO is often considered the propeller head; they have a deep technology background. • Their approach is one of defense in depth. They understand the technology and controls that are in place. • They focus on new cyber products and transforming defenses by adding new functionality. • The technical CISO prides themselves on their up-to-date knowledge and has a strong external network.	• Naturally, the technical CISO cares less about business engagement and stakeholders. They find them particularly stressful and may avoid these interactions. • This CISO would also not typically enjoy board and management, as they don't speak the same tech language and it would take more effort to explain their plans.
Transformational CISO	• The transformational CISO is very similar to their CIO cousin; they strongly lead change and uplift their environment at pace. • A strong vision and courage are mandatory for this CISO.	• This CISO is a person who creates significant stress in their enterprise by leading disruptive change. This may also lead to stress for their CIO as their own digital initiatives may be impacted by these changes. • They will typically drive hard deadlines, particularly those that have been set by regulators. • There is also stress in keeping up with the latest cybersecurity threats and trends.

CISO Styles	Approach	Stress Indicators
Operational CISO	• The operational CISO is focused on ensuring that IT systems are running smoothly and efficiently. • They are typically experts in IT infrastructure and operations. • Operational CISOs are often responsible for managing IT budgets and resources and ensuring that IT services meet the needs of the business. • Strong process thinking is mandatory for the operational CISO.	• This CISO runs the BAU of cyber operations systems to ensure that these technology components run efficiently. • These CISOs may experience more stress if they are not given the resources they need to do their job effectively.
Business CISO	• The business CISO is similar to the strategic CIO. They have deep links with the business and have a strong focus on strategic alignment. • Due to stakeholder engagement, they are often at the table and invited to discussions on the future business roadmap. • They are trusted to present to the board a good understanding of business impacts and risks.	• This CISO will feel pressure to balance their own security objectives with other business priorities. • At times, they will be stressed by the actual and perceived lack of support from the business for cybersecurity initiatives. • The CISO may have difficulty communicating how actual security risks compare to the trade-off of stopping other business activities.

CISO Styles	Approach	Stress Indicators
Partnering CISO	• The partnering CISO is a similar construct to the partnering CIO. There is an absolute focus on relationships with stakeholders and business units. • The partnering CISO spends their days meeting with business stakeholders, external partners, and their staff. • They are naturally aligned to have constructive discussions and proactive engagements.	• This CISO also works with stakeholders and must listen to their requirements while balancing their own needs. • This style of CISO has reasonable levels of stress as they have to balance the needs of the stakeholders with the needs of their own team. • Their greatest stress comes when they have conflicting requirements and must resolve this.
Risk culture CISO	• The risk culture CISO really cares about building a strong cyber risk culture. This means that the CISO leads by example in *setting the tone*, both internally and with external parties. • This CISO wants all team members to take personal accountability. They also want their board to be well informed and set the tone for cybersecurity culture for the enterprise.	• This CISO has a long-term goal and will at times find it stressful to measure their short-term impact. • There will be stress due to the scale of their business and the fact that staff turnover will imply that new cyber training is required as new DNA enters the enterprise.

Table 8.2 – CISO stresses by persona type

Regardless of whether you are a CIO or a CISO, you will need to figure out how to work effectively with your counterpart. As we have explored, how you operate and how the other party operates have to be catered for. When you have a similar approach, it might be a match made in heaven, or it might lead to conflict as you think *too* similarly.

It is a useful trait to be observant and reflect on how best to interoperate. This can then in return help you to reduce the stress that you face in this relationship.

Now that we have looked at the life of both a CIO and CISO, let's explore a few ways in which CIOs and CISOs can manage stress effectively.

How the CIO and CISO manage stress

While you can never make stress fully disappear, to survive, the CIO and CISO must be fully aware of the strains of the position. There is a need to be resilient and spring back into shape to avoid burnout and breakdowns.

Should you not have your own plans to manage this stress, it may get the better of you and shorten your tenure even further. When you consider how you match up the CIO to the CISO, this may provide you with some insights into stress management. Consider drawing an arrow between yourself and your counterpart, whether they are the CIO or the CISO. Being of a similar persona to your counterpart may lead to more harmony, but it could also lead to you competing for the same resources.

While not overanalyzing this, some simple mapping can help you both to find your common ground and potentially assist you to partner better together.

Managing stress – How do you and your counterpart match?

Figure 8.4 – How you align with your CIO/CISO

Each CIO/CISO will have a natural style and approach. These personas are what you bring to your job every day, and this directly impacts how you operate in the role. Also, you must work effectively with your counterpart, who may also be your direct manager.

Your predecessor will have had a certain persona; you will then come to the table and adopt their strategy but with your own approach, which is likely to be different to theirs. Regardless of whether you are a CIO or CISO, you will have to match up with your counterpart and make the relationship work.

There is no one wrong or right answer – there just needs to be as much unison and synergy as possible. The key point is that there will be some natural tension and stress points in the relationship; how you operate to make things work is key. This CIO-CISO dynamic will create some level of stress that you must accept. Having now understood the importance of effective stress management, let's figure out a few ways to see how we can better manage this:

- **Acquire the best talent**: Firstly, the CIO and CISO must surround themselves with the best talent available. They need experienced colleagues who are not afraid to take on challenges or provide strong counsel to their boss.

  I've seen some CIOs and CISOs that are perhaps threatened by having too strong team members who could challenge their own position. In my view, you are better set up for success by taking on a strong team to help you more effectively manage your agenda.

- **Have a support network**: While direct subordinates can provide a support network, there are also external colleagues who can act as sounding boards to give guidance and a different perspective. I've been the chair of the FS ISAC Regional Strategic Committee and have built a personal network through various speaker engagements.

  The network requires *give and take*, and it is always refreshing to realize that you aren't alone and that there are other CISOs and CIOs with the exact same problem as you to resolve. So many times, I've noted that similarities are more common than differences and that it is only the logo of the entity that is different.

- **Adopt a healthy lifestyle**: Personally, I have practiced yoga and other forms of training. In my case, practicing yoga every day has helped me clear my mind and get strong clarity of my day's objective. Mindfulness training can be done through walking, running, swimming, and other forms of exercise. I have also taken my own team through mental resilience training, where we explore the neuroscience of the brain to understand better how it works.

- **Maintain a healthy work-life balance**: As the CIO or CISO, you will need to take on full responsibility for your role without taking yourself too seriously. Taking a strong stance on managing your own resilience will help you to avoid burnout and enable you to better handle the pressures of the job.

  Each person will need to explore how to maintain an acceptable work-life balance for themselves. Some individuals can find peace and mindfulness in different activities than just sports. Just going for a walk or reading can promote your own separation from work and allow you to naturally decompress. Some may get this release from playing with their kids or pets. It really doesn't matter, but you must find your own retreat and be able to de-stress by reflecting on what is more important.

  This absolute focus (*Drishti*) is the key to being able to prioritize what really matters at the present moment and being aware that a few items might need to be left off the list for the time being.

We are individuals first and leaders second, and the ability to self-manage your stress is an essential competency. Your team needs to be able to identify their approach to doing so. They won't be of much value to you if they are not coping themselves.

Summary

A day in the life of a CIO or CISO is full of pressure and stress. This stress comes with the role, but rather than just accepting it, there should be approaches that you take to help manage the stress effectively. That said, the pressures are different and there are distinctive degrees of stress with each CIO and CISO persona.

Not every CIO/CISO is the same, and their natural styles will bring different pressures and strains. Some very specific forms of stress were explored in this chapter, and these will vary depending on what kind of CIO and CISO you are. There are different CISO and CIO approaches that will have distinct stress indicators.

Noting your natural approach as a CIO or CISO will help you reflect on how to operate. Making the relationship with your CIO or CISO counterpart work effectively can certainly help you to alleviate stress levels.

Getting strength from your network and community is another option to manage your stress, as they understand what you are going through. The role can be lonely, with few to talk to, and this can be a personal burden. Therefore, you must personally explore how to manage this stress through your own work-life balance.

In the next chapter, we will start to explore survival skills for the CIO and CISO. These survival skills will assist in your own stress management but also provide you with approaches to help you with the sustainability of your career as a CIO or CISO.

9

CIO and CISO Survival Skills

The American reality TV show *Survivor* is the inspiration for this chapter. In this series, a group of diverse strangers are castaways on a remote island or setting. They must fend for themselves, covering all of Maslow's hierarchy of needs – air, water, food, shelter, sleep, clothing, and so on (Note that I will be referring to this as Maslow's theory). Likewise, in a corporate setting, the CIO and CISO will need to adapt and figure out their own survival. Yes, you are on top of the pyramid and in the lead, but to retain this position in the long term will involve survival skills. Unlike other senior roles and your peers, there are higher and perhaps unreasonable expectations of this position.

The complication with CISOs is they are required to protect an enterprise; however, they do not control every facet of that organization, including staff behavior, user-defined systems, and adoption of **Software as a Service (SaaS)** providers. Similarly, the CIO has this remit, plus the requirement to build an IT strategy to drive innovation, business processes, and data within the enterprise. Again, the CIO does not directly control all IT and technology processes that operates within the company, and the SaaS technology trend only makes this worse.

As you can imagine, the lives of the CIO and CISO are not for the faint-hearted. These are difficult roles that come with the territory. There will be stresses and strain, which will last for the position's duration and not be confined to the first few years. In terms of industry averages, the CIO and CISO are usually in their role for two to three years. Accordingly, the CIO and CISO will need to develop their survival skills to stay in their jobs and have some career longevity. These survival skills will be explored in this chapter, and mastery of them will come from practice and strong commitment.

The fundamental question that is to be addressed in this chapter is as follows:

What are the survival skills for the CIO and CISO?

In this chapter, we will discuss the following topics:

- Exploring Maslow's theory in the context of CIOs and CISOs

- Building a strong foundation

- Cultivating skills to ensure longevity

Exploring Maslow's theory in the context of CIOs and CISOs

Maslow's theory proposes that humans are motivated to fulfill needs in a specific order. The hierarchy is depicted as a pyramid, with the most basic needs at the bottom and the most difficult needs at the top, as shown in the following diagram.

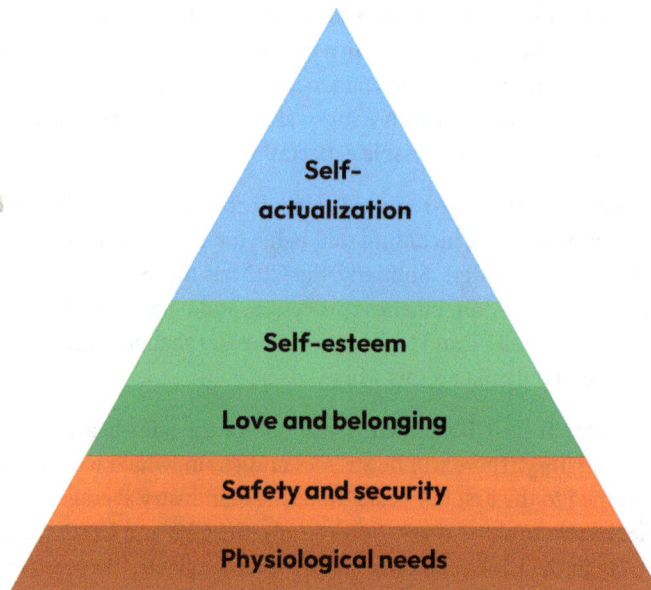

Figure 9.1 – Maslow's hierarchy of needs

The five levels of Maslow's hierarchy of needs are as follows:

- **Physiological needs**: These are the most basic human needs (e.g., food, water, clothing, and shelter)

- **Safety**: This includes security, stability, and protection from harm

- **Belonging**: This includes intimacy, friendship, and a sense of belonging to a group

- **Esteem**: This includes a sense of achievement, self-respect, and confidence
- **Self-actualization**: The highest level, with the need to live a meaningful life and reach your full potential

This is a well-understood and accepted model; however, it is not often applied from the CIO and CISO viewpoints.

An understanding of Maslow's theory can help the CIO and CISO to have an appreciation for their own roles and why they feel a certain way. It will certainly assist with their own personal resilience, allowing them to assess what is behind any moment when they feel they need to "fight, flee, or freeze."

The model will also help them develop empathy for their teams, which will certainly also feel these fight, flee, or freeze responses. And, as the leader strives to build a safe environment, the team should also strive to get to the belonging, self-esteem and self-actualization level.

In the following table, we explore how Maslow's theory applies to CIOs and CISOs:

Maslow's need	CIO and CISO context	Current concerns
Physiological needs	• CIOs and CISOs require the basic need to feel that they have a comfortable and safe work environment. Their role is challenging, and they will need some time to rest and recharge.	• The CIO and CISO both have high-stress roles that often will feel like they engage them 24/7/365. It is not uncommon to take calls and meetings during holidays; unfortunately, projects and incidents don't always stop during annual leave.
Safety needs	• CIOs and CISOs need to have job security and personal financial stability. They have a basic need to feel secure in their work environment.	• The CIO and CISO are both high-profile individuals who work on strategically important initiatives. Their peers and management will themselves feel some pressure, which usually gravitates to the CIO and CISO. • They are also often personally targeted by external hackers. Thus, they will not find it easy to feel safe at work or in the home environment.

Maslow's need	CIO and CISO context	Current concerns
Love and belonging needs	• CIOs and CISOs desire to feel like they are part of a broader management team and that their contributions are respected. • They will also want to have constructive relationships with Peers, Management and the Board.	• The CIO is usually a direct report of the CEO and therefore part of the management team. However, the CISO can operate often at a level below.
Esteem needs	• CIOs and CISOs will want to be valued by their Peers, Management and the Board. • Moreover, to make a positive strategic contribution to the enterprise.	• Often the strategic programs of the CIO and CISO are not always aligned with the priorities of Peers. Hence there will be some tension of resource and priority conflict. • Their progress is also often not celebrated as the work is infinite and never really completed.
Self-actualization needs	• CIOs and CISOs both need to feel like they are making a significant difference to the mission. • They both want to believe that they are operating at a peak level and driving significant change.	• For the CIO and CISO, driving transformation is never trivial and is significantly complex. • Unfortunately, transformations are often problematic and require personal commitments to be sacrificed to achieve the outcomes desired.

Table 9.1 – How Maslow's theory applies to the CIO and CISO

I would suggest that CIOs and CISOs should reflect and understand Maslow's hierarchy of needs as a framework to cultivate better work settings, for themselves and their teams. This approach can also enable them to develop their own human resource strategies to engage and motivate staff.

Maslow's theory can be used to help develop aspiring CIOs and CISOs. This same approach can help you consider how your peers and colleagues themselves may be feeling, as they are subject to the same pressures. Your fellow stakeholders must manage their own stress and strive to achieve a work-life balance while looking to achieve a broader mission to make a difference.

In the next section, we explore some aspects of job choice for different CIO and CISO roles. Is your role one that will help you maximize your own self-worth? If we apply and use Maslow's framework, can we reach the top part of the pyramid??

Building a strong foundation

Given the difficult career that a CIO or CISO will face, it is imperative that they have a strong foundation. These foundations are based upon choosing the right role and at the right time. Life is usually about making the right choices, and this is even more so in the case of the CIO and CISO roles. In this section, we also will explore in further detail stakeholder management and evaluate a broader set of peers and external parties.

Making the right career choices for yourself

During your career as a CIO or CISO, there will be moments that will define you and what you want to stand for. These are moments of truth, where you have choices to make in terms of your career direction, and each of them may be significantly different from the other.

Choosing *job happiness* is what I recommend. The phrase *job satisfaction* is one that I don't use. The word *satisfy* to me is not strong enough to convey the joy that your role should bring to you personally. Each of us should be jumping out of bed with energy and enthusiasm for what the new day can bring. Having a passion to get results and to advance the cause is the right focus.

For me, job happiness means that my role brings a sense of worth and accomplishment (self-esteem). I measure this by my confidence in the value I bring and how my own competence matches this role.

Then, if I'm extremely fortunate, job happiness may also bring some degree of self-actualization – I perceive that I'm fulfilling my own potential and personal growth. This job as a CIO or CISO helps me to become and embody an ideal leader. This may materialize as a feeling that you get occasionally when you are at the top of your own profession, can anticipate how the technological world is progressing, and then take proactive actions. There is no better feeling as a leader.

Therefore, I challenge you to ask yourself the question: "What is the right job that will take me to this level of self-esteem?"

What is the right job?

This is a key question, and it must be very clear in your own mind what the key criteria are that allow you to accept a role and become excited about it. In my opinion, there should be three criteria to ascertain this. Let's explore them:

- **An exciting challenge that allows personal growth**: Do the CIO and CISO roles not only have the right combination of playing to your strengths but also have sufficient "stretch" that will help you grow further? What excites you about this role? What about the job takes you outside of your comfort zone, allowing you to grow?

  This may be broader business or geographical scope. It might be the complexity of an organization or the transformation that is required. For some, this may be the technology that is used or will be used by the company. Either way, you can personally relate to this challenge, and it motivates you.

  For the CIO or CISO, this challenge could be the scope of the budget and resources and how critical the roles of technology and cybersecurity are to business success. For the CIO operating in a digitally enabled enterprise, this makes the transformation impact greater, and the cybersecurity stakes will increase for the CISO.

- **An industry and location that is aligned with your personal values**: Does this new CIO or CISO role operate in an *industry* that you have a passion for and aligned with what you see as valued? By valued, I mean that the role makes a difference in your eyes. For each of us, that definition is very personal, and this alignment, therefore, has some individual significance.

  A good example is one career choice I made to join Eli Lilly and embark on a CIO career across many countries and regions. The pharmaceutical industry (and, in this case, Eli Lilly) provides therapeutic treatment for diabetes, mental health, cancer, and other important human diseases. My alignment to this cause is reinforced by the strategic long-term nature of the industry, with products going through a 10–20-year development life cycle. I valued this and wanted to do my part to help accelerate this innovation. As I have had family members who have or could use these products, it just felt aligned with my beliefs and, therefore, felt right.

  The *location* of a role is also so critical. Taking on the right role in the wrong location can cause family stress and issues. Most executives are successful because they can get balance in their lives by having great support on the home front. This holds true for all genders. It is possible to try to ignore this factor in the short term, but no doubt the stress created by not working in a location that you are happy with is simply not worth it.

- **Having an overall compensation that matches your expectations**: Note that this is listed last. This is on purpose, as without the previous two bullets ticked off, it matters little if the compensation matches your expectations. To me, money and financial compensation, indeed, are important but should never be the main reason you take on a new CIO or CISO role.

  If you are unfulfilled by the challenge of a position or feel somehow compromised by working in an industry, then that won't change. Money can't buy happiness – it's true.

  In this IT industry, CIO and CISO positions are in high demand, and as a result, there has been some spiral effect on salaries and bonuses. This can lead to unreasonable expectations, thereby overshadowing the previous two criteria.

  My advice is that, for the CIO or CISO role, explore the range of compensation that you can receive. Being a good "match" for a role means you are in the "acceptable" range of candidates. It would be better to be at the highest end of suitable candidates, but just make sure you are not at the lowest end.

While securing a challenging job you enjoy with a competitive salary in a desirable location is crucial, long-term success in the demanding role of a CIO or CISO hinges on another key factor – understanding your stakeholders.

Let's shift gears and explore how to recalibrate your stakeholder analysis. This was important in your first 90 days and is even more important for your long-term survival in the CIO or CISO role.

Recalibrating your stakeholder analysis

In my first 90 days as a CIO and CISO, I completed a Myers-Briggs analysis with my new team. That is a useful starting point, as you now know much better how your stakeholders think and can also start to get a clear picture of how they interoperate with each other.

Again, the approach I use is quite simple and can be completed on a whiteboard with colleagues in the privacy of a meeting room. Note that the stakeholder analysis is designed to help you identify any new stakeholders and reassess older individuals and groups, who have a strong interest in your domain.

Analyzing your stakeholders is a tool for CIOs and CISOs to ensure alignment of their projects and initiatives with business needs. By doing so, they will have a sense of what level of support they will receive from key stakeholders.

In truth, some of you may have probably assessed some stakeholders in your own head. But it is always valuable to try to engage your team to understand where each internal and external stakeholder is plotted and have these expectations shifted.

There will be new projects and initiatives that come into play that weren't yet considered when you first started in your role.

Also, in your role, you must liaise with a broader group of stakeholders, which can go beyond what was superficially completed before. The stakeholder analysis looks outside of the company and includes the board. By going broader, you may want to consider some of the stakeholders that are in bold:

- Employees

- Senior management

- Colleagues in other IT departments

- **The board**

- **Suppliers and partners**

- **External customers**

- **Regulators**

- **Government**

- **Industry colleagues**

This process of stakeholder engagement will assist the CIO and CISO in managing the expectations more effectively. The broader group of stakeholders can often provide you with anecdotal stories and feedback that you can share with management and the board, from whom there will always be a concern about lagging behind competitors.

Having a sense of these benchmarks can provide some level of comfort, as technology as a topic can confuse many as being too "techy". A simple comparison to another industry or a positive comment from a regulator can provide good ammunition for the CIO and CISO when there are doubts expressed about the maturity of an IT or cyber control environment.

Next, let's learn how we can analyze stakeholders using the Power Interest Grid.

Stakeholder analysis through the Power Interest Grid

The Power Interest Grid, as shown in *Figure 9.2*, looks to segment stakeholder types into one of four quadrants. These include the following:

- **High power and high interest**: These are the stakeholders you must manage closely, as they are key decision-makers who can help you succeed.

- **High power and low interest**: You must keep them satisfied, but they have little energy to help you succeed. They do, however, possess power that can hurt you, if they choose to exercise this.

- **Low power and high interest**: These people need to be kept informed. They want to help but lack the power to be useful, to be honest.

- **Low power and low interest**: Don't over-invest in this category, as they don't have any stake or power to influence your direction.

Stakeholder analysis – Power Interest Grid

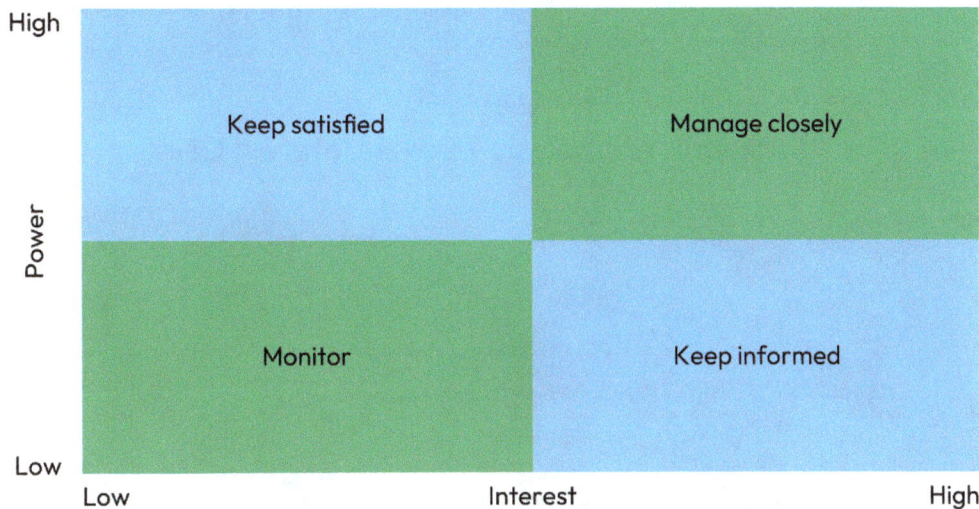

Figure 9.2 – The Power Interest Grid

The whole purpose of the exercise is to understand and assess the power and influence of relative stakeholders in the work ecosystem. Once this is assessed, you should develop tailored plans to engage with them, including communications. In some cases, they will need to be involved in the plans for business and system contingency. Again, the degree of investment will depend on where they appear in the grid.

By involving your own team in this exercise, there will be a stronger sense of joint ownership of this plan, as you, the CIO or CISO, are not able to address any actions by yourself.

By plotting your key stakeholders, both internal and external, you can gain some insights into where the potential pinch points can exist. There are stakeholders that will require more attention and have to be **managed closely**, while some others you can **keep satisfied**.

This level of granularity will be a good exercise to anticipate the power and interest dynamics and, therefore, the stakeholder areas that you as the CIO or CISO should devote more or less time to improve your role longevity. Here is a summary of this exercise:

- **Manage closely**: This would include the CEO and your regulator. They both have a strong interest in your work and are not afraid to exercise their power.

- **Keep satisfied**: This would include the COO, CFO, and other key internal stakeholders. They can exert their power to hurt you, but they don't want to do your job for you.

- **Keep informed**: This would include your third-party partners who have a strong interest in your engagement; however, they don't exert much power to hurt you.

- **Monitor**: This includes some peers and commodity vendors.

In *Figure 9.3*, I have plotted these stakeholders across a grid.

Stakeholder analysis – Power Interest Grid

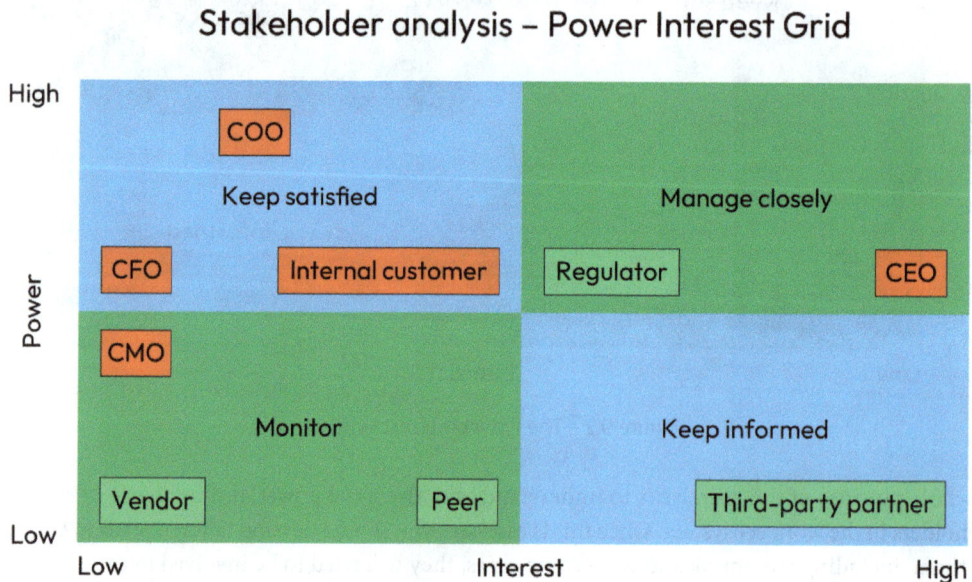

Figure 9.3 – Stakeholders mapped to the Power Interest Grid

One key thing that you will quickly notice is that it is impossible to keep everyone happy. Indeed, the CIO and CISO are often the people who deliver bad news that no one wants to hear. There are many examples of shared information that is unpopular. Here are a few:

- Key systems will have to go offline to allow for security patching

- There has been a cyber or data breach, and your transformation project must be prioritized after a new compliance remediation

- A third-party supplier has been compromised and some data has been lost

- Key resources must be switched to another mandated initiative

- Your technology is obsolete, and the cost to support it will increase dramatically

- The budgets have been cut and your project is impacted

As the CIO or CISO, this is where the development of your "soft skills" in navigating through any potential land mines will be critical. I've seen and had to deal with these examples and many others. These are never pleasant, and the conversation is unavoidable.

We've established the importance of understanding your stakeholders for long-term success. Now, let's address the reality of the average CIO/CISO tenure. While the numbers might suggest a revolving door, achieving longevity is very possible. Here's what sets those who thrive apart.

Cultivating skills to ensure longevity

We have already noted that the average tenure for a CIO and CISO is two to three years, and while there are some exceptions of executives who can enjoy real longevity in their roles, these are not the norm.

There is a private joke I have heard about CIOs and CISOs, where the role acronyms are said to stand for "*Career Is Over*" and "*Career is Shortly Over*," respectively. For the average CIO and CISO, longevity in their roles is not typical. Therefore, while this is said in lighthearted jest, there is a reality that these positions are intrinsically demanding. To survive in the longer term, you need to have a strategy that is flexible and can pivot across different dimensions. What I mean here is that the CIO and CISO should seek broader experience across a variety of industries, different scales of organizations, and diverse business and technology challenges. Thus, their career is a series of traditional, lateral, and spiral career shifts.

Longevity for the CIO and CISO should be viewed differently from the colleague who has worked in the same enterprise for 20–30 years and worked their way up in an organization. This model is more common in countries such as Japan, where staff have been trained as generalists and promoted into areas that are not their area of expertise.

While the number of years of service is not necessarily a good or bad attribute, there are some financial attractions to remaining in a role, to allow stock options to accrue and be realized. From my own experience, I was a CIO at one single enterprise for a 15-year stretch of my career; however, this was as a CIO across five different countries, with different business operation scales and strategic importance.

The key was that I had both career and personal growth in these CIO assignments, allowing me to feel that I attained the top level of Maslow's pyramid across each role. My longevity and ability to move between CIO roles was the result but not the objective.

In every CIO and CISO career move that I have personally made, this has been accomplished without bringing old colleagues with me. Instead, I've worked with and developed the team that I have inherited. While I'm not judging here and understand why this often occurs, sometimes the CIO and CISO are faced with the challenge of not being sure who to trust. Given this scenario, they will often turn to a few trusted friends who can help them accelerate their change agenda.

This is a common survival approach that CIOs and CISOs have employed to help them gain some longevity. Think about this – they are at a new company, often without any allies and with significantly challenging deliverables. Hence, it can be a natural impulse to turn to old colleagues.

However, be aware that there is a flip side to this. Although a leader bringing their friends with them can accelerate short-term progress, it may breed some distrust, as the existing team at your new company may be in the so-called "storming" phase.

By honing the right survival skills, you can not only weather any initial storms but also establish yourself as a leader who thrives in a dynamic environment. Let's explore some other essential skills that will equip you for long-term success, such as building alliances.

Building strategic alliances

On the TV reality show *Survivor*, the best competitor is not always the physically strongest. There is a strategy for surviving, and that means forming alliances to give yourself the best chance to be the winner. The winners will often methodically work with other competitors if the alliance makes sense.

In the world of the CIO and CISO, they are often in the firing line. It is not uncommon for the incumbent to suffer from burnout from work pressure. Their roles are themselves under pressure from many internal and external stakeholders.

Take my own example – in my last two positions in two different roles, I have had six different supervisors. That works out to be an average of a new supervisor every year over the last six years. This means that continually building alliances has been a necessity.

Moreover, my key stakeholders also have churned with market changes, so they too are partnerships that get reset due to changes in circumstances. Your ability to build relationships will help you get the job done – it is a simple truth. You don't need to have everyone as your friend and ally, but you don't want to have enemies everywhere.

The alliance partner for the CISO

When you are a CISO, it is often the case that the CIO is your supervisor. In larger enterprises, there are often many CIOs that are your peers and stakeholders. But are they your natural allies? From my experience, some CIOs get cybersecurity and are fully committed to supporting the cause. However, I've found, to my own dismay, that the majority may get cybersecurity but show ambivalence to the cause.

Many peers will view cyber as a negative area, focused on serious incidents and career-threatening ramifications. Cybersecurity also competes with the CIO's agenda for resources and funding. As a former CIO who was extremely committed to cyber, I found that my enthusiasm was not met by my peers.

In my view, true digital transformation is all about delivering the outcome of an improved customer experience. If a digital experience is impacted by a cyber event, then it doesn't matter how funky and innovative the technology is, as it no longer provides a positive customer experience This is why I was motivated to gravitate to a CISO role after more than 20 years in my capacity as a CIO.

The alliance partner for the CIO

Conversely, when you are a CIO, who can you confide in? Yes, you can get counsel from the CISO, but sometimes, that may not be feasible, given the confidential or sensitive nature of the discussion. There may be some senior peers that you can talk with in confidence, but again, that will depend on your own comfort level with these parties.

What I learned was that the COO and business unit leaders are often very concerned about technology-driven transformation, the disruptive impact of cybersecurity, and how this will impact their desired operations. For their own survival, they have (in some cases) concluded that they should build a strong alliance with the CIO.

Should the CIO look further afield for support? And, for that matter, should the CISO also consider having some external third-party engagement?

Where else to look?

Each CIO and CISO is often able to find both intellectual and moral support from their external peers. There are many working groups, industry forums, and conferences that can cultivate a coalition to learn and share. By doing so, you may be able to gain some valuable informal and anecdotal benchmarks on digital transformation or cyber controls from other peers.

This itself may help with your survival by providing an alternative perspective on a specific problem that you alone have. These collaborative friendships can help each CIO and CISO in the following ways:

- Sharing technology and cyber best practices from others in the industry

- Promoting industry standards for new developments such as the cloud, digital, and cyber

- Cyber intelligence sharing can help reduce overall risks in a larger ecosystem

- Providing tips on who else to talk to who may be able to provide some guidance

These examples can provide you with additional insights that will be valued in your role as CIO or CISO, and perhaps by your boss!

Finding a mentor

There may also be some grounds to consider finding a mentor or personal coach. Such an arrangement can be valuable to get a sounding board and receive some independent coaching. The job of the CIO or CISO can be quite isolating occasionally, and you may want to discuss an emerging issue or problem. At times, these can be quite confidential and sensitive.

The value of an independent mentor is that they can give you a different perspective. At times, just having someone senior to listen to your concerns will help you think through a "live" problem. Let's consider some of the reasons to consider engaging a mentor:

- **To learn from their "battle scars"**: The mentor could be a former CIO or CISO who has relevant experiences that are similar to what you face.

- **A network of networks**: The mentor will have a different network of contacts that you can learn from and gain career opportunities.

- **Developing leadership skills**: The mentor can provide their expertise to teach you and help you reflect on areas to improve.

- **Problem solving**: The mentor can assist you with a difficult project or situation. They have an objective perspective that can be extremely valuable.

A question that you may have is, how do I find the right mentor to help me? This can be done through your own internal or external network. This may also be one area to discuss with your own boss, seeking their endorsement and acknowledgment that you are trying to address some areas for improvement.

By referring to your own SKEB assessment to see what attributes you seek to build, you can sync this with how your mentor assists you in providing some alternative insights.

One very good area of improvement to consider is becoming more maze-savvy and being able to operate in your enterprise more easily.

Effectively managing political situations

This chapter has focused on sharing some survival skills for the CIO and CISO. Part of this survival requires being aware that there are fellow peers, managers, and stakeholders you will encounter that are extremely "political" and, therefore, difficult to handle.

I would never want to be seen as "political," and nor do I propose that this is ever the best approach. Rather, I advocate that you should develop your political savviness. This means you should be able to identify when others are being political and figure out how you can deal with this.

As a CIO or CISO, there are decisions that you make every day that could upset someone else and be the cause of some political retaliation.

We discussed that you must understand the business landscape with your stakeholder analysis. This helps to identify the key stakeholders and what their motivations and personal agendas are. These parties have contrary objectives that can impact your efforts. In these circumstances, we explored allies and found those who have common interests that can support your position.

However, when you are faced with a political situation, then what? The following suggestions are some principles that you can consider to help you avoid politics. Let's remember that politics will always be present; what you can control is how you act in response to them.

Stakeholder analysis - Power Interest Grid

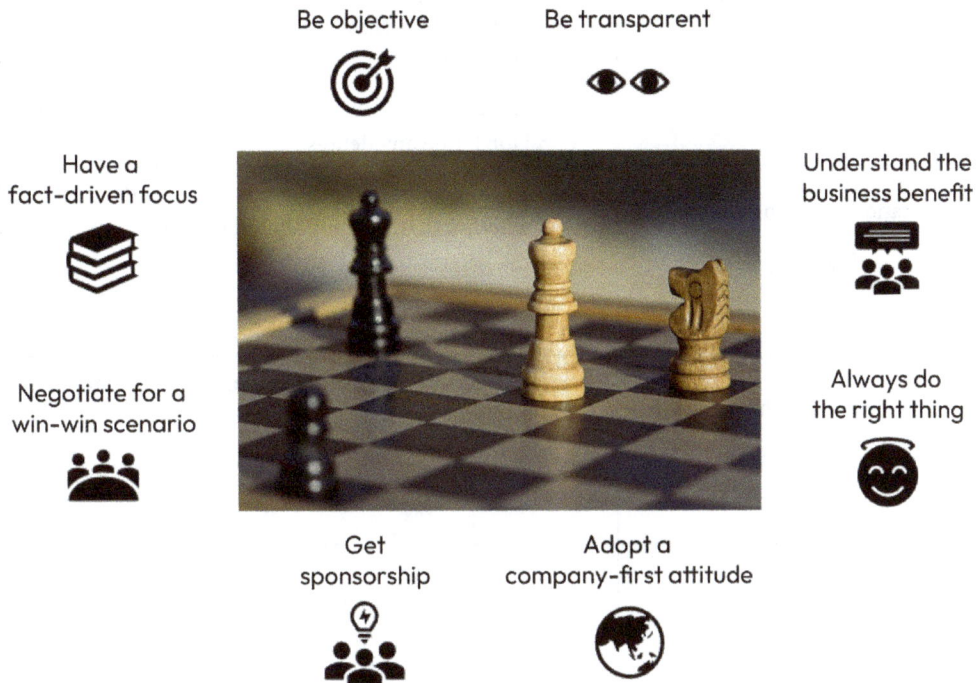

Be objective

Be transparent

Have a
fact-driven focus

Understand the
business benefit

Negotiate for a
win-win scenario

Always do
the right thing

Get
sponsorship

Adopt a
company-first attitude

Figure 9.4 – Principles to avoid politics

Some common-sense approaches can be adopted to help you navigate through political waters. Here are eight simple things that you can keep in mind to provide you with a platform that is not seen as "political," and this way, perhaps you can avoid politics.

These are explained further in the following table:

An approach to avoid politics	Description
Have a fact-driven focus	• Focus on the facts and what data exists to support your recommendations. • Try to avoid being personal or advocating personal opinions.

An approach to avoid politics	Description
Be transparent	• Be able to pass the "red face test" (meaning that you don't get embarrassed and turn "red," as you aren't lying) • Communicate openly in a two-way conversation. Don't hide any details.
Understand the business benefit	• Explore what the greater good is to understand whether there are any shared interests across the parties.
Get sponsorship	• Get buy-in from senior leadership who can advocate for your initiatives and navigate political hurdles.
Negotiate for a win-win scenario	• Learn to negotiate effectively, address concerns diplomatically, and build trust with diverse stakeholders.
Be objective	• Stay neutral and avoid taking sides. • Remain professional and focus on the merits of an issue, regardless of personal alliances or disagreements.
Adopt a company-first attitude	• Make decisions based on what benefits an entire organization, not individual agendas or personal gain.
Always do the right thing	• Be a role model to do what is right. • Ask challenging questions, but don't launch personal attacks.

Table 9.2 – An approach to manage political situations

As the CIO or CISO, you must navigate political behavior. There is no way to avoid managing these situations, and this will require you to exercise discretion and tolerance. This is part of your job – perhaps not the most enjoyable component but necessary to be successful.

Your role also requires that you influence stakeholders and external parties, such as regulators. This will require you to build trust and strong relationships to tackle political challenges.

Here are a few more points to consider when managing political situation:

- Join the dots of a situation yourself so that you can understand hidden agendas
- Understand what is considered acceptable behavior
- Don't lower your values when you see others do this (take the high road)
- Use your political savviness to achieve positive outcomes for an organization
- Focus on your and other people's accountability

Once you have mastered the art of being politically savvy in tricky situations, you need to stay ahead of the curve by continually investing in yourself. In the following section, we will see how to do that.

Maintaining continuous growth

Your long-term career as a CIO and CISO requires your commitment to be a life-long learner. This growth can polish your rougher edges and make you the best version of yourself. No CIO or CISO starts as the finished product; we must be open to learning and development, which means being receptive to feedback.

It is often said that feedback is a "gift". I did resist believing this during my career, and frankly, it took some time to understand that feedback is neither right nor wrong. It is purely a perspective from one of your stakeholders and staff that you can simply choose to ignore or instead reflect and consider.

Yes, it is true that feedback can hurt and make you feel uncomfortable. I would recommend that each CIO and CISO become their own role model in giving and receiving feedback. This takes practice to get better at both ends of the transaction.

In my own career example, I've learned from my role models – bosses, peers, and staff, and not necessarily in that order. We see and observe good and bad examples of Skills, Knowledge, Experience, and Behavior. It then requires your own focus to decide what to emulate and what to avoid copying.

My own approach has been to lean on my personal strengths in the following areas:

- Strategic thinking
- Communications
- Business orientation
- Being a quick learner

These attributes have enabled me to move between companies, countries, and industries. It is often said that "*what got you here won't take you there.*" There is truth in this statement. However, I would say that, in my experience, what got you here will take you there, but you need to keep developing your strengths and not rest on your laurels.

Yes, it takes some degree of *courage* to do this. However, I think this is the best approach, and it maximizes the fun that you can have with your career. Note that I have again put a specific focus on honing your own soft skills. For the CIO and CISO to survive, they are expected to be great across all their dimensions – people, process, and technology. This means not only having a strong knowledge of technology but also being skilled at stakeholder management and managing your staff.

A well-rounded CIO or CISO is what is on your potential employer's list of desired characteristics. They want this rather than individual brilliance in just one of these dimensions.

Summary

You have invested significant effort to become a CIO and CISO. There will need to be an ongoing investment to remain and prosper in this role. In this chapter, we explored how Maslow's hierarchy of needs applies to the CIO and CISO. Each of you should aspire to attain the highest levels of the pyramid.

The chapter also provided some guidance on the following:

- Building a strong foundation
- Redoing your stakeholder analysis
- Building partnership alliances
- Getting a mentor or coach
- Navigate yourself when others start to be "political"

You've got this!

Your long-term career as a CIO or CISO will require an ongoing investment to develop yourself. The role will continue to be dynamic and shift as technological change drives uplift. As the leader overseeing this shift, you must grow with this trend. To resist will not be an option. Your goal is to be successful and not just survive, and to do so will require careful stakeholder management, building strong alliances, and acquiring mentors, both inside and outside of your company.

There will come a time when you, for different reasons, must change roles and look for an alternative path. These are difficult decisions to make and execute. I've learned during my own career that "seeking that next elevator" requires some practice. I have met many industry colleagues who have been long-serving executives for 10–20+ years at a company, but they lack the "muscle memory" of applying for another job to easily secure their next role. This next chapter is my guide to helping you in this stage of your journey.

Part 4:
What's Next in Your Career?

In this final part, you will learn about career management – when is the right time to seek alternative paths or another CIO or CISO position? Or should you consider consulting? Your overall life and career path is one to be mapped and you should always try to think two jobs ahead. Your career may pivot from one CIO role to another or from CIO to CISO (and vice versa). In *Chapter 11*, we will discuss an alternative role as a Risk Management Executive. This path is not taken often but should be considered. A final bonus chapter will explore what many CIOs and CISOs do in their retirement. Several rewarding post-career moves are explained, and this is when they don't want to sit on a beach!

This part has the following chapters:

- *Chapter 10, Looking for the Next Elevator*
- *Chapter 11, Risk Management as a Career Option*
- *Chapter 12, What CIOs and CISOs Do in Retirement*

Looking for the Next Elevator

I recall waiting for an elevator in a hotel lobby in Sydney once. As the doors opened, the waiting individuals started to shuffle in, and the music that was playing became louder and more upbeat. It was indeed a party elevator, and not only did it take you up to the restaurant level but it was fun as it detected that there was a small crowd onboard! Moving to a new job in search of better opportunities may be likened to a staircase or an escalator – both travel upward, but the elevator has a higher velocity. We want the next elevator to take you to a new, fun challenge.

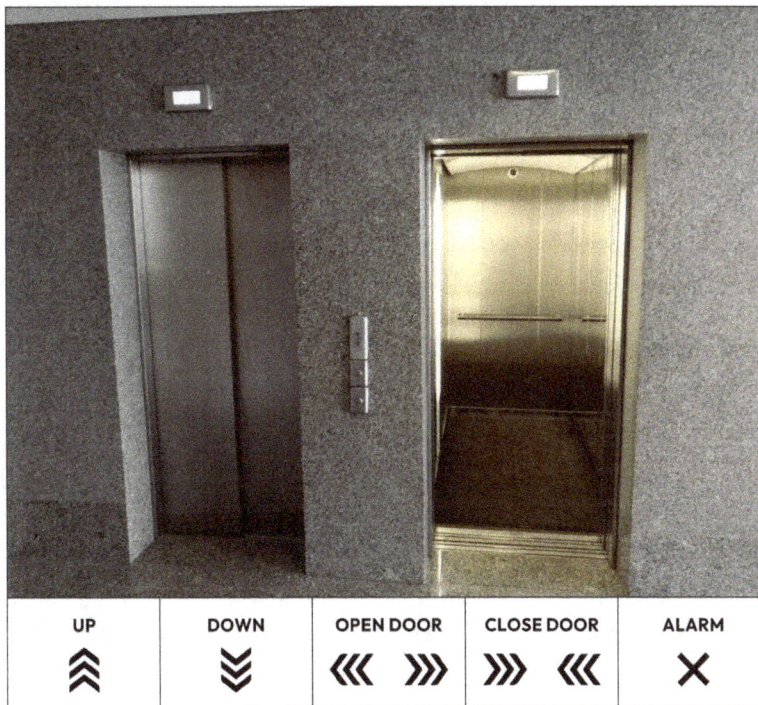

Figure 10.1 – An elevator as a metaphor for your career
(source: https://commons.wikimedia.org/wiki/File:Elevator_2.jpg, uploaded by Misiokk, used according to the Creative Commons Attribution-Share Alike 3.0 Unported license)

You should have a good personal sense of when you need to change and seek the next move. An elevator, like a gondola cab, takes you both upward and downward. This journey is certainly faster than taking the stairs and you can move quickly in either direction. This chapter talks about being in this situation and looking for a change.

> **The fundamental question that is to be addressed in this chapter is as follows:**
> How do I plan for my next CIO or CISO role?

What should you think about as you consider taking the elevator, what are the alternative paths, and how do you regard your current position and the next in regard to your overall career and life plan? Is there a transit lounge option that you can consider? And once you've decided to move, how do you hold the door open for your successor?

We will be covering the following topics in this chapter:

- Why look for the elevator?
- Choosing the next elevator
- The transit lounge
- The elevator to building your career portfolio
- Holding the door open for your successor

Why look for the elevator?

Sometimes, it doesn't matter how hard you have tried to succeed or fit into an organization, it is just not going to be a good match. I've seen individuals try to struggle with this dilemma and it can be painful to watch. We all know how to ride an elevator; it is simple, right? But making career decisions is not that easy. The following is a guide to pressing the elevator buttons:

UP ⌃⌃	We all naturally want to progress upward. There may be career life choices that you make in which the elevator is not the right mode of travel and the staircase is a better (slower but more appropriate) choice. Note that being overly ambitious can also be perceived as a career derailer, so be careful not to push too hard.
DOWN ⌄⌄	Again, we may have career life choices in which we choose a different CIO or CISO role that may be perceived as inferior than the current one. This may be to escape a bad culture, a poor boss, or both. There might be moments when poor health or familial commitments would be good reasons to take this approach. You can take the stairs to step down to a lower level.

OPEN DOOR ⟪⟫	Taking the position that you are *open* to movement is the first step and is often completed in confidence with a select group of colleagues and recruiters. This is like saying you are not looking for a new role but are open to understanding the market.
CLOSE DOOR ⟫⟪	Conversely, taking the position that you are *closed* to movement and not interested in changing roles can be a subtle approach to pretend to be less interested as, typically, not being available can make you more desirable.
ALARM ✕	*Let me out now! I hate my current job and must leave!* This is an extreme case and should be used with caution. Even in circumstances where your situation is valid, this can backfire, and you could be seen as a poor performer with part-ownership of the problem.

Table 10.1 – Elevator buttons

The question is when do you press the button and which one should you select? We will look at a few situations that will accelerate your walk to the elevator to take an alternative path to the one you are on.

To do a postmortem can be a good idea, but in truth, you should already have a good idea of why you want to move. Traditionally, we have seen evidence that CIOs and CISOs will change roles every three to four years, but we all know that statistics can lie and be misleading.

Clearly, while this rule of thumb doesn't always follow a strict timeline, there are reasons why this occurs and each of these may have some level of interplay:

- **Burnout and stress**: Earlier in this book, I detailed how difficult these senior positions can be. With this complexity and constant pressure comes stress. Frankly, the roles are thankless and rarely is your work and effort appreciated, as it is akin to being on a treadmill that is constantly moving.

  Some CIOs and CISOs need time and space to reset themselves within a new industry or different scope role to give them a mental break from the burden they will face daily. The word *burnout* is a strong one; however, this is real, and stress-related absences occur at all levels including some senior CIOs and CISOs.

- **Seeking new challenges**: Once you feel that you have put in the hard yards (for example, completing a large transformation or successful remediation of a cyber regulatory gap), then your current role may be less motivating for a CIO or CISO who likes the sense of challenge that comes from a new position. Naturally, roles sometimes attract these *adrenalin junkies* who may be attracted to the next raging fire that has new difficult problems to address.

Both the CIO and CISO roles are serious roles that tend to be very demanding and complex. Plus, they require the incumbent to keep learning and changing with the business evolution that occurs. A new job change can feel like a literal breath of fresh air, as there is no backlog, and starting fresh can feel simply liberating.

- **The boss and the culture**: It is easy for there to be misalignment between your boss and yourself, and more broadly, for the company culture to be wearing on you. What makes a company great is the culture, but this can be a double-edged sword and annoy you most.

  The CIO and CISO require a risk culture to be lived and role-modeled by all their stakeholders and the board. When this risk culture is not aligned, it makes the role much more difficult and, indeed, somewhat impossible to succeed.

  In some cases, this may be acceptable as you enjoy the challenge of the position and complexity to get to the end outcome, but if this shifts or continues to not be anchored, then this can be a major frustration. In that case, it absolutely makes sense to do an active search for a new position.

- **Finding better paying opportunities**: The global cybersecurity gap of more than 4+ million professionals is true for every level. This is no different for the CISO and CIO roles, which are in extreme demand, and salary packages can be accelerated by switching roles. There are some downsides, as longer-term retention share options and grants could be in jeopardy and sacrificed if shifting roles.

  There is also the other side of money. Many CIOs and CISOs have joined a role to affect a transformation change. However, when the company has re-prioritized, the adrenalin fix that comes with the position may disappear, leaving the senior executive feeling somewhat perplexed. They are valued for their change transformation competency, but this is now on hold. Such scenarios can be frustrating for the CIO and CISO to remain in the role. The key frustration is that while their mission has not changed, the funding may mean that this journey has been extended longer than expected. In short, it is not what they signed up for.

Now that we have acknowledged the realities that can lead you to think of a career change, let's turn our attention to a proactive approach. Before diving head-first into the plethora of new opportunities, there are some key considerations to solidify your decision and prepare for a successful transition.

Choosing the next elevator

It is best not to rush into any decision and jump at the first moving opportunity. As said before, an elevator is a device to assist people in traveling between levels. Indeed, this may mean that you can travel both upward and downward. As the master of this decision, it then becomes a purposeful one around the criteria to choose to help you make this move.

If you want to change roles, I would encourage you to ask yourself these questions first:

- **What would excite me?**

  As you want this elevator to take you somewhere that is fun, it is critical that this new role truly excites you. Therefore, you should ask yourself a series of questions to be sure:

  - What development areas would I want to develop? (See *Chapter 2*, referring to **Skills, Knowledge, Experience, and Behavior (SKEB)**)

  - Does this imply any specific new emerging technology – AI, data, cloud, or mobile – that I would like greater experience with?

  - What about a new industry that I have not worked in previously? Or one that I would like to return to as I sense that the elevator is moving fast?

  - What about a change of country and a new lifestyle?

  - Would this new role be a great move for my family?

  - Have I considered who I would really like to work for? Is there a person that I admire and want to learn from?

- **Have I scanned the horizon completely?**

  Your objective should be to have full visibility of all the alternative options in the market. Hence, you must actively network broadly while being discreet and purposeful. Ask yourself these questions:

  - Have I looked more broadly at megatrends and things happening in the last year? (Often there are many new developments that you don't have time to study.)

  - What networking have I done in terms of connecting with my peers?

  - Have I spoken to my mentor to seek their advice?

  - Have I attended technology CIO and CISO events to expand my network?

  - Which recruitment and search firms have I had a casual chat with?

  - How is the local market for senior roles that would be of interest?

- **Have I revisited my career objectives?**

  This is a pivotal career moment, and you should reflect on where you want to go when selecting the next elevator. You must start with your own career objectives. Ask yourself the following:

  - Should I circle back to review my career and position objectives? (See *Chapter 3*.)

  - Do I need to revise one or both objectives?

  - Given my personal circumstances, do I want to be more aggressive or more conservative with my career plans?

- What is my personal ambition in terms of a larger role, a larger team, a larger budget, or a larger company? Do I want to reduce the risks in my career?

- Do I just want a stable life?

- Am I now more driven to contribute more to society?

- **Have I considered potential new roles?**

  When it comes to potential new roles, you must complete your own due diligence to understand whether something would be a good fit for your new career objective. Ask yourself the following:

  - Is the new role in a company that is growing and expanding?

  - What don't I know about this new role that worries me?

  - Do I know someone at the targeted company? Or do I have someone in my network that I can be introduced to?

  - Do I understand their culture and mission? Is that a good fit for me?

  - Who would I like to be working for? What is that person's brand?

  - What does their LinkedIn profile tell me about them? Are there any red flags?

- **Have I dug deeper into the new role?**

  Now that you have identified a good opportunity, how do you determine whether this is the right one for you? How much digging and further research must you complete? Start with these questions:

  - What does the job description say and not say?

  - Why did the last incumbent depart?

  - Is the job doable? Or is it too challenging?

  - How stable are my direct reports? Has there been a churn?

  - Will the compensation be in the ballpark range that I seek?

  - What do I think of the peers that I have met in the process?

- **How do I decide?**

This new opportunity looks interesting and worthy of consideration – these are always exciting times for you to contemplate and reflect. Remember that moving to a new CIO or CISO job is not something that you *must* do. So how do you ensure that you make the right decision? You must ask yourself the following:

- What attracts me most to this opportunity?
- What part of the offer don't I like?
- Does the offer put me in the ballpark of the salary package range I want?
- Does it feel like a good personal fit for me in terms of the job challenge and industry outlook?
- Do I have any reservations about my new supervisor?
- Can I envision being successful in this role?

In the end, this must be your own decision, and you can make any decision work well provided you have considered the options and weighed up the upsides and downsides. You will have your own gut feeling that will guide you. In my view, your instincts are usually correct, and I've always trusted mine while making a difficult career choice.

My advice is to take a blank sheet of paper and write down the pros and cons. It will always surprise you that there is more noise swilling around your head than there are issues to resolve. Once you put this down in simple points, then the decision can be more easily made.

Your mind will be full of comparing different options and scenarios. When you have more than one offer, this can make the decision even more problematic and challenging for you. Let me share an example where an important career decision was difficult to process and conclude. In this case, I sat on this decision for two or three months with different offers in hand.

A Japanese Zen garden in Kyoto – Ryōan-ji – holds a special place in my heart for making my first CIO career decision. I had been offered three executive roles – one was to become a full partner at Ernst & Young at the age of 34. They had offered me a signing bonus, an MBA, and a study tour, along with accelerating the process to have me conferred within a month, rather than the normal process. I had another partner offer from AT Kearney, another consulting firm, and a CIO role with Eli Lilly.

I sat in the Zen garden with my wife and three children. They watched the garden and imagined islands, turtles, and crocodiles while I contemplated this career choice. All I could see was the rocks, and how my career roles led me to a larger one from smaller and medium-sized ones. For me, this symbolized the fact that my next career decision would be a significant one that had longevity. This made me decide to change my mind and take the opportunity at Eli Lilly.

In my mind, there were salary packages and bonuses along with notions of prestige and titles. In the end, while I thought these were important, they were never the deciding factors.

Figure 10.2 – Ryōan-ji in Kyoto, Japan

My first role as a CIO was decided this way. I had, of course, discussed this many times with my wife and with mentors. As I said, in the end, this must be your decision.

It was one of the best life and career decisions that I made as it launched me into a career that spanned living internationally in five cities across four countries. This decision allowed me the privilege to travel as a tourist to 52 countries and experience many moments that took my breath away.

The impact of a great decision is that it will open doors to new possibilities for your life and career.

The transit lounge

There are times in your career when the opportunities just don't match with what you are seeking. As a leader with personal financial responsibilities along with professional pride, it may be hard to contemplate why you aren't snapped up immediately in a new position.

The truth is that the hiring process is extremely complex – usually with some science but, moreover, more human choice. I've seen interviewers make irrational decisions when selecting a new employee, and that is usually due to self-interest overshadowing what is required for a senior role.

There are times when you may challenge what other stakeholders want in the position. This then may lead you to consider your alternatives to bridge this situation. In my career, twice, I have used consulting as such a conduit. I have had stints with Ernst & Young and KPMG Consulting; in both cases, there were different reasons why I ended up taking this route.

What I would say is that the professional services path is a great transit lounge, if you want to consider this option. First, the opportunity will provide you with a broader perspective to provide service across many clients and industries. This will force you to learn new industries and processes.

Along the path as a senior CIO and CISO, you will also sharpen your selling skills, which is a major expectation at the salaried partner level. This will usually involve some adjustment for some executives as the sales cycle, while also sometimes required in large enterprises and companies, will be every day. Another by-product will be that your own presentation skills will be enhanced; by that, I mean the actual performance art aspect of delivering a pitch. But also, you will want to craft your message with more innovative and creative resonance.

The partnership model is not for everybody, though. I would advise you to only enter this model with a positive attitude toward being a full equity partner. Yes, this may be a transit lounge, but you can't think of this or let others know that this is how you consider this stage.

Privately, you will also find that there are significant differences in how these consulting firms operate from that of an average enterprise or company. There will be frustration in some of the model's weaknesses that you will discover, so I suggest you understand this but focus on the strengths.

What are the strengths? These consulting firms are structured to organize agilely to present the best position and pitch for work. The resources are always stretched, which is not dissimilar to your old role; the significant difference is that there will be more junior staff assigned that don't have the same depth of competency.

As an experienced hired leader in these situations, you will be expected to step up and be the subject matter expert, and in technology, this can be a stretch. Hence, the adage of *being comfortable with being uncomfortable* still holds true, perhaps just as strong in the consulting world. For the CIO or CISO who has moved into consulting, they may find that sales and the targets that come with the role are the largest adjustments.

While you may not enjoy sales and selling, the beauty of being a hunter is that you can enjoy the spoils when you are successful. While it is not for everyone, the other consideration is that you can reposition yourself to gain experience in other industries. That could be another great rationale for taking on consulting: to reinvent yourself.

The transit lounge will take you somewhere and the question is how you use this time to evaluate options. Yes, you will be in a high-profile role to meet other executives who are your potential customers, and this can itself lead to a role.

In my own case, I was offered my first CIO role at Eli Lilly when I was the Acting CIO there for a few months. This was a difficult decision for me as a full equity partnership was being offered to me.

This then becomes a personal decision around what you value and want.

However, the transit lounge can also be a frustrating place if your next flight is delayed and you can't then find your next seat. Consulting work can be very rewarding financially and perhaps also provide job satisfaction. Earlier in my career, I did get a little frustrated that I could devise a new IT strategy or digital transformation strategy – then as a consultant, not see this delivered to the end outcome.

That was my own personal rationale for going down the CIO path where I had the opportunity to do this from end to end – from strategy to implementation. It is good to be able to test yourself that you are indeed delivering real value and benefit, and not just making promises that you have no accountability for.

Later in my career, I did have a few years consulting for myself under my own name, and through a few small boutique technology firms. At a more senior level, I provided strategic review and advisory for CIOs and CISOs, which I found very rewarding. This, again, was a transit lounge but without all the frills and trimmings.

The knowledge and experience that I have gained during my CIO and CISO career were then used to advise others on more specific transformations. If you like, this is wisdom as I was providing guidance based on my own successes and mistakes. In a long career, you will collect these battle scars, and this will form what becomes your portfolio.

Leveling up to build your career portfolio

We have used the elevator to get up to new levels of being a CIO and CISO. You must enjoy the journey along the way and consider the overall destination of where this elevator is taking you.

The elevator is a means to an end and not *the* end. Your own career and life are what this is all about. This journey can take you to some destinations that were never envisaged when you boarded on the ground floor.

The elevator can take you up and down. It sometimes moves slowly between levels and at other times it can travel quickly. When you decide to get off, then there are *Open* and *Close door* buttons that you can press. There is also an alarm button that you only use in emergencies when your elevator is not moving anywhere, and you want to get off. Like the elevator, your own career is a series of these movements between levels, which, over time, will form a portfolio of your experiences. Your career portfolio is like a good list of investments; you can gain strong wisdom from those choices that did well and from those that didn't when you take a step back and look at your portfolio view.

Some of your portfolio will do better for you than other parts. Each of us can take a risk view of this portfolio and invest with different strategies.

When you put all your investment eggs in one basket, then it may do well, or it may only do okay. A portfolio implies that you are spreading your risks more broadly by using several strategies – for example, in the personal investment case: shares, property, cash, and bonds.

Your goal is to win in life, and that doesn't mean you are gambling and taking bets. That may be a tactic that is considered but it is not a long-term way to win. When it comes to investments, though, information is key. While it is not possible to predict the stock market, it is true that investment bankers still have not been disintermediated by technology (yet). That is because information flow is not equal, and investment bankers can gain some insights from their market sources as to where the market is moving.

In our career example, you need to gain these same insights and ensure that you gain wisdom to help you individually win. Your career to date has been a series of jobs and roles that have been stitched together where you have gained new SKEB along the way.

These challenges of success and sometimes failure are all learning experiences. The extreme cases always provide you with both good and bad examples to draw on. In your career, you will have faced problems that required strategic decisions and different tactics/strategies. The experiences you gain will provide you with the fabric for your own career.

> **Key career tip**
> A rewarding career is a series of roles in which you are learning from your successes and failures.

The everyday life of navigating and being resourceful as a CIO and CISO will build your own confidence in what you can do and provide a real-time assessment of what makes you uncomfortable. It is not only what you experience as a leader; there will be lessons gained from seeing others in action and the mistakes and triumphs that they make.

Being a lifelong learner is what I have strived to be. Some of you may be naturally more intelligent from an academic or emotional intelligence standpoint. But I'm not talking about the starting point – this is a journey, and lessons are learned along the way.

The hardest journeys that are uncomfortable can often be the most memorable and rewarding. In contrast, the 5-star hotels that anticipate all your needs and pamper you are indeed lovely but are not what I define as a necessity for a good trip. When I reflect on my many travels to destinations, there are many that have the attributes of "uncomfortable" that I have thoroughly enjoyed.

Allow me to reflect on my own career and how I built this using a portfolio approach – a career that I have already discussed in *Chapter 3* as having different models: traditional, linear, transitory, spiral, and so on. For some, the traditional career ladder is one step at a time to climb up. However, that approach wasn't for me and the absolute joy I had was to take a journey in my search for continual learning and to have fun!

You may recall in *Chapter 5* that I developed my own set of expectations, and this was collected on my journey:

- Bring your best version of yourself to deliver value (every day)

- Be comfortable being uncomfortable

- Have a hunger to keep learning new things

- Work as a team

- Have fun!

I have collected this set of expectations over many years. It was never just a list that I started with, but when I started CIO and CISO roles, I wanted to be able to share this as "My Expectations."

By using this approach, I have collected a portfolio of life experiences that looks like this:

Author, board advisor, Buddhist, CIO, CISO, change agent, consultant, director, father, Fintech advisor, grandfather, golfer, international expatriate, investor, keynote speaker, magazine contributor, MBA, PC engineer, risk leader, rugby fan, student, traveler, transformation leader, wine enthusiast, Venture Capital partner, and yoga student.

You will notice that many of these items are not work-related. Indeed, you cannot separate yourself artificially; to be authentic, it is much easier to bring your whole self rather than thinking only of the part of you that is shown in the public light.

My own journey has been about living life and collecting experiences that make me a better person and leader. The journey will mean that you build relationships along the way that are lifelong and help reinforce your own purpose. Giving back is part of this loop and, in a small way, this book is part of my own endeavor to help your career.

The following is a visual that I prepared of my career and life portfolio:

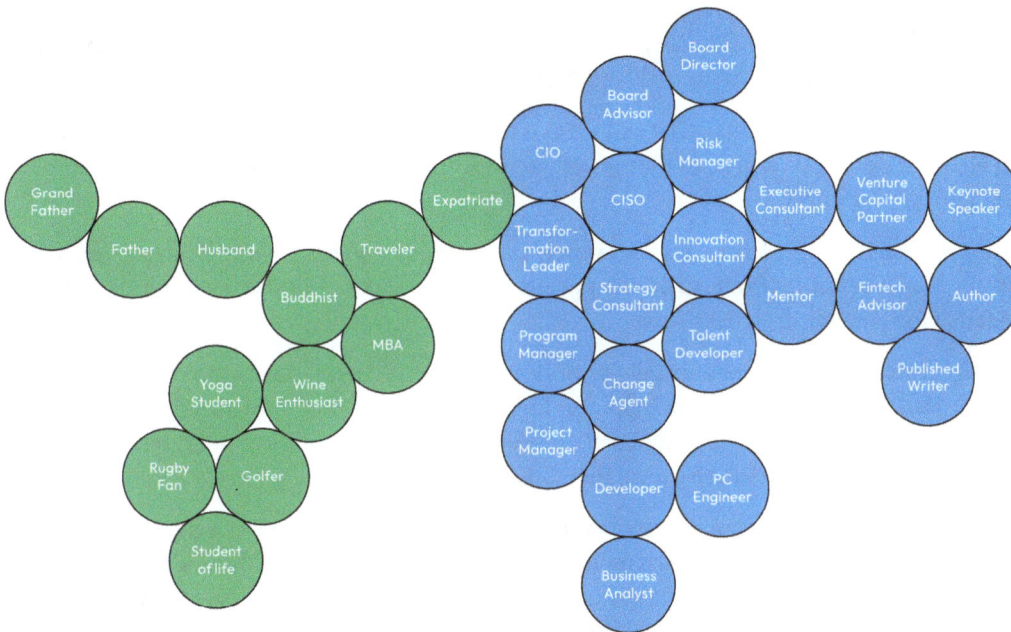

Figure 10.3 – Career and life portfolio

Figure 10.3 shows how I have managed and owned my career. Using the principles I defined, this was used to decide directions. Rather than a simple sequence of next roles, I have managed a career across multiple industries, multiple countries, and different traditional paths.

In my life, each experience has a connection to another experience. The more personal ones are in green and then the professional-related ones are in blue.

The career portfolio map that you see here shows how your career is built from these experiences – for instance, the links between expatriate experience with some of the senior CISO and CIO roles that I have successfully undertaken.

Let's remember that this started out traditionally but then you make choices that serve to grow your portfolio and the new skills, knowledge, experience, and behaviors can then be used in any role that you perform. I would encourage you to try to develop your own career and life portfolio.

If I consider my own example, I talked in *Chapter 1* about understanding your own brand. For instance, strategic thinking was one of my strengths that I always felt was a superpower. This led to opportunities to lead large-scale transformation exercises and be a change agent. I learned how to be a good change agent and always want to drive change, so regardless of whether my role is a consultant, CIO, CISO, or risk manager, I will have a natural default to doing this.

This brand that you build is unique to you and will power your career. Thus, my coaching to you is to embrace the diversity of your career and life choices. Be bold to make choices that are not customary and think outside of the box. This will give you a unique selling proposition and you will then stand out from other candidates in terms of the value that you can bring.

Being a standard garden variety of CIO and CISO is fine and there is nothing wrong with going down this path. What must be clear is that the portfolio returns are potentially much narrower as there are fewer options on the table to explore.

Besides, your ability to have fun will be severely hampered as you are placing stronger limits on yourself. The more fun your career is, the more you will do better work in that environment. That is the kind of fun elevator that I'm seeking.

In my own career, having the opportunity to work as a CIO, CISO, and later, as a risk executive was never the plan. It just felt like the right opportunity at the right time with the company and industry (elevator) that I wanted to jump onto. Each of these career moves requires careful reflection and perhaps some mentoring to test your own thinking.

Indeed, I'm certain that for more than half of my career and life portfolio, I now look back and say "Wow, I did that!" There are backstories to each milestone – for instance, taking on expatriate assignments is a significant shift that requires preparation for yourself in language and culture training and for your family in terms of relocation, new schools, friends, shops, and jobs for your partner, plus, of course, language training.

In my case, it was as simple as my wife once remarking, "Wouldn't it be nice to work overseas?" At the time, I did not work for a multinational company and no transfer option was on the table. It just became a thought that ended up being part of a later career choice and where this opportunity was highlighted as a potential future scenario.

I would note that moving from country to country is difficult but incredibly rewarding as you just learn so much, which also means it is extremely tiring. Similarly, to move from one industry or sector to another is also not at all insignificant. Typically, there are barriers that HR search firms will put up and if you don't have previous experience in ABC or XYZ, your CV will be considered but it is hard to get off the long list.

That's where your career portfolio will give you the confidence to tackle these large professional shifts and make this successful. Good luck with building out your career path and portfolio. There will always be some blessings required, but hard work and some reflection on what strengths you bring will give you the best chance to have a great career.

Holding the door open for your successor

You are about to travel on the elevator up to another CIO or CISO position: however, there is still the courtesy of holding the door open for the next person. That person is your successor to your now old job.

As a CIO and CISO, you often don't have the benefit of receiving a handover. The prior incumbent may have already departed or did not leave on good terms. This can make your role more difficult. Thus, I want to talk about how you should arrange your own handover.

I can recall receiving some limited handover and it was clear that the prior incumbent did not want to surrender the role. Hence, they were not very forthcoming with sharing any important information, nor were they willing to spend time to help me succeed.

You should want your old team to prosper and do well when you depart; therefore, taking time to plan your exit and assist the successor is a critical part of your brand.

To prepare for this handover transition, it is important to first develop a transition timeline and plan for how you can help the new CIO and CISO succeed. During times of transition, there will be a sense of loss by your old team as they worry that their efforts during the year won't be recognized or understood.

Figure 10.4 – Holding the door open for your successor

(source: https://www.flickr.com/photos/141761303@N08/36756252356, uploaded by www.amtec. us.com, used according to the Creative Commons Attribution-Share Alike 3.0 Unported license)

Your key stakeholders will be concerned about unfinished projects and commitments you have made, and whether these will be honored by your successor.

The first step for you is to develop and collect handover documentation. Just imagine what you would like to tell your younger self if you were the new CIO or CISO. This can include formalized aspects of your role and informal rituals and customs.

Documentation should include the following:

- IT and cyber strategy
- Portfolio and program management plans
- Organizational architecture (org charts, capability assessments, job descriptions, etc.)
- Staff assessments (objectives and performance and salary reviews)
- Operational documentation (process and procedure documents, manuals, policies, etc.)
- Secrets management (master passwords, encryption keys, etc.)
- Vendor contracts
- Stakeholder strategy and notes

A handover can take from 1 week to 6 months; it just depends on the circumstances. The longer durations I've seen occur when the incumbent is retiring or has taken up another role in the organization. There is no right length, and it is likely that you will also have unanswered questions.

The documentation noted previously can be supplemented by sharing what you had in your own 90-day plan and suggesting that they use this as a template.

There will always be a little tension between the prior incumbent and the new CIO or CISO. It is just natural that the new CIO or CISO will want to make their own mark, and that there will be some defensiveness if there is any criticism.

Some tips I would suggest to help are as follows:

- **Don't rush**: You didn't get to where you are in a few weeks, thus it is important to allow some "soak time" for all the details to make sense. If your successor is coming from outside as an external hire, then there is much more to learn. Don't try to get through all your transition materials in one sitting or one day, as that is sub-optimal.

- **Treat the new CIO or CISO as you would like to be treated**: The newly hired CIO or CISO will probably make some mistakes, and it is not your place to make them feel bad about this. Providing context of what happened and where this was when you started in the role is helpful, but more important is to help the new CIO or CISO build their own perspective on the forward strategy.

Ideally, you can suggest some unsolicited advice on where more attention could be given. The trick is to not communicate in a fashion that patronizes but advise that this was a theme that you had identified that was on the backlog and, hence, not yet addressed. It is more of something to keep in mind when you reset the IT and cyber priorities in the future.

- **Celebrate the past**: Be sure to highlight to the new CIO or CISO the victories that the team has accomplished and do this publicly so that your old staff feel they get their credit. Being generous with praise does not cost you anything and will help the team as they deal with their own insecurities.

  Having a few joint handover sessions with your lead team will enable this to happen organically and without appearing to be concocted.

- **Shadowing**: The best way to hand over is to schedule time for you and your successor to work together in meetings and workshops. This will allow you to introduce the new CIO or CISO to your stakeholders and to do this in the context of the project team meeting or forum.

  The discussion is then relevant and specific, rather than being just the introduction 1:1 meeting. Shadowing is a very hands-on approach and can accelerate this handover.

  A few tips to make this work well are listed as follows:

  - Block out your schedule to enable shadowing

  - Be sure to brief on what is expected to be the mood and content of each meeting

  - Later, be sure to also debrief on what did happen

  - Feedback if there are other areas that you identify would benefit from the shadowing approach

- **Follow-up**: Finally, offer to provide further details after the transition is over. While you may be leaving this role and moving on to a new company or role, there would still be much pride in seeing your team succeed in the future.

Be gracious and that is good karma that comes back to you.

Summary

You're looking to change your role and seek a new challenge. This may be due to burnout and stress or because your boss and the culture do not match what you're looking for. It could also just be a commercial decision to earn more compensation.

How do you choose your next step? What is exciting to consider? And how should you start to scan the horizon for these opportunities? This may mean that you revisit your career objectives and then evaluate these positions and then make a considered decision. I have shared my Kyoto Zen garden story as an example to reflect on making such a choice when there are options available.

The transit lounge in consulting is often an alternative path. There are tricks and tips to go down this route. Evaluate your career as a portfolio and see how your overall self-development is synergistic and how everything you learn will enable further options for growth and opportunity.

Your career choices over the long run will define you. Each choice that you make will enable you to gain new experiences that grow how you operate. These form what I have called your life and career portfolio – this is a set of experiences that become your expertise.

Some of these paths will lead to further growth and others may have fewer long-term advantages than you can realize.

Now that you have chosen a new role to move to, you have an obligation to provide a great handover to your successor. How you do this will be important and mirror how your reputation in terms of the deep footprints left behind will be remembered. Your handover is the reverse image of your first 90-day plan, and it should be completed thoughtfully and with grace.

The next chapter is an optional one that I added. In this book, I have encouraged you to consider a career as a CIO and CISO. Note that I did not use the word *or* here, as it is my recommendation that you consider both paths up the mountain. In my career, I have learned skills, knowledge, experience, and behavior that have allowed me to move between industries and countries, plus across C-suite roles.

My addition of *Chapter 11* is to allow you to also consider technology risk management as an extra career option. This certainly was not one that I considered early in my career, but it made more sense as I wanted to think two jobs ahead and consider non-executive board positions.

11

Risk Management as a Career Option

This chapter will explore an alternative career option that the CIO and CISO most likely have not considered, and you may have discounted this idea as well. My intention in this chapter is to provide you with an alternative perspective and open this path to you as an option.

You will have an opportunity to consider a new trend to have stronger lines of defense across first- and second-line risk functions. This is being driven by cyber-attacks across all industries and increased regulatory scrutiny.

Risk Management is something that the CIO and CISO do every day in their day-to-day role. They must decide where to focus and where to ignore for now. This is a very dynamic process that increasingly requires technical depth. For this reason, the CIO and CISO have battle scars from their experience allowing them to deeply understand technology, cybersecurity, and data, making them ideally suited to consider this alternate career path.

> **The fundamental question that is to be addressed in this chapter is as follows:**
> Why should I consider Risk Management as a potential career path?

This is another path up the mountain that should be evaluated. It is not typical, but that is exactly why you must give it some serious thought. In this chapter, we will be covering the following topics:

- Why Risk Management is a viable option
- Why might you want to cross over?
- Risk Management as coaching
- Finding your way to become a coach

Why Risk Management is a viable option

Throughout my career as a CIO and CISO, I have worked with Risk Management and alongside them as a partner. I had felt that Risk Managers were people who had never actually performed the CIO and CISO role that I performed but somehow had the responsibility to comment on my work.

Then, later in my career, I started working with Risk Managers who had been a CISO or a CIO. This made me reconsider how their skills, knowledge, experience, and behavior made them excellent in this role.

I never seriously considered Risk Management as a career option, and subconsciously I had not really considered this path. But, in this chapter, let me explain why I pivoted and changed my own thoughts on this, and how the skills, knowledge, experience, and behavior that you have obtained in the CIO and CISO roles are perfect training to be a risk leader.

As cybersecurity continues to be the number 1 risk for enterprises, the focus is on ensuring that all lines of defense are strengthened. Thus, there will be a greater drive to have more technology-savvy Risk Management Executives.

Risk Management as a discipline has evolved as technological advancements have been made in society. It has become more formalized and developed with regulatory requirements pushing for proactive Risk Management.

In many enterprises, Risk Managers have been treated as a necessary stakeholder who may slow you down but who you must work with to manage technology and cyber risks as a CIO and CISO. It is also true that Risk Management manifests differently across various companies, and this variation means that it is not always clear what the purpose of this function is. I recall in CIO and CISO roles that there were 2LOD Cyber Risk roles, Operational Risk roles, Control office roles, Quality Assurance, among others.

The accountability buck stops with the CIO or the CISO. When you are a CIO or CISO, you will be empowered to take leadership and drive change. Unfortunately, it is very easy to see how Risk Management becomes the blocker of this process and adds barriers to change.

Risk Management professionals, however, are not alike and come from diverse backgrounds of work and study. They are often strong in problem-solving and have good analytical skills. Their education is regularly accounting and finance, although there are Risk Managers who have sundry miscellaneous work experience.

A typical Risk Management **Skills, Knowledge, Experience, and Behavior (SKEB)** profile looks like this:

	Classical Risk Managers - Summary
Skills	• Problem-solving skills • Analytical data skills • Stakeholder management skills
Knowledge	• **Certified Information Systems Auditor (CISA)** • **Certified Risk Manager (CRM)**
Experience	• Broad business experience • Compliance and regulatory background • Project leader experience • Law, finance, or accounting degree
Behavior	• Tenacity • Personal resilience • Look at issues top-down

Table 11.1 – SKEB for Risk Managers

In this section, I outlined the classic profile of Risk Managers we have all worked with. Many CIOs and CISOs already have these SKEBs. But any of you who have served as a CIO or CISO may perhaps look at this and still not be convinced. However, I am seeing a new trend, and the traditional role is shifting with new requirements. These new requirements are changing the status quo.

As I reflect on my own career, it is fair to say that each of my career changes was opportunistic, but I also recognize that I wanted to build a unique brand that would add more value to my CV. Just consider this for a moment: if you have already been a CIO multiple times, then yes, you can repeat or contemplate taking a new CIO or CISO role in a different industry, company, or country. Why not also consider the Risk Management Executive position when you are contemplating a career move?

The next section explores how more CIOs and CISOs might want to consider this alternative path to becoming a Risk Management Executive.

Why might you want to cross over?

To start, I would argue that CIOs and CISOs are Risk Managers anyway. They don't like to think of themselves in that vein, but this is a fact. The CIO and CISO want to maximize the value they can deliver with new projects that reduce the residual risk.

Many CIOs operate in dynamic transformational environments, often viewing the scenario they face as a situation where the glass is full or half-filled. A typical CIO can see the downside of a particular situation, but given their agenda for digital transformation, they must look at the upside more than the downside. On the other hand, the CISO can typically view the world with a slightly more negative lens and view the scenario as a situation where the glass is half-filled or empty. The Risk Management Executive should ideally be more similar to the CISO.

With that said, there are indeed many attributes that the CIO and CISO have that can be applied to a professional career as a Risk Management Executive. These include the following:

- Taking a risk view of any situation to make decisions under time pressure
- Managing tech and cyber controls
- Stakeholder management
- Relevant technology background

It is also true that there are more instances of CIOs moving to CISO roles and vice versa rather than moving to Risk Management. Conversely, CIOs or CISOs moving into Risk Management Executive roles is still not a typical path, but I believe that this trend will continue to grow. My sense is that the role of Risk Management has increased stakes because boards and regulators are getting more concerned about the maturity of cyber and tech resilience.

Therefore, Risk Managers have to also step up with more authority in an increasingly complex technology environment. The new requirements are for deeper technical skills and know-how, including an understanding of new technology such as the cloud, AI, data, networking, and engineering. This is a very different profile from the traditional model. In my view, the new-age Risk Management SKEB looks more like this:

	New Age Risk Managers – Summary
Skills	<ul><li>Hands-on technology skills</li><li>Advanced data analytics skills</li><li>Problem-solving skills</li><li>Analytical data skills</li><li>Stakeholder management skills</li></ul>

	New Age Risk Managers – Summary
Knowledge	• **Certified Cloud Security Professional (CSSP)** • **Certified Information Systems Security Professional (CISSP)** • **Certified Information Security Manager (CISM)** • **Certified Information Systems Auditor (CISA)** • **Certified Risk Manager (CRM)**
Experience	• **Cyber, technology, data, ML/AI degree** • **Deep technology business integration experience** • **Project delivery** • **Engineering**
Behavior	• **Leadership** • **Change agent** • Tenacity • Personal resilience • Look at issues top-down

Table 11.2 – SKEB for new-age Risk Management Executives

It should be clear that the profile of new requirements (shown in bold) in the preceding table indicates a significant shift in upskilling existing incumbents or bringing in new options that match these criteria. Thus, the CIO and CISO will be seen as viable options to take on this challenge.

In fact, in my case, when I went to find my successor for my Risk Management Executive role there was a shortlist of five that included four CISOs and one more traditional Risk Manager. Yes, I had a personal bias to replace myself with a CISO with technical depth and all the stakeholder management skills to succeed.

My experience is that all executive roles, such as CIO, CISO, and Risk Management Executive, come with stress and challenges. It is not the case that you have to be burnt out to want to change to Risk Management. The level of tension and anxiety is clearly different in all of these roles.

How each person manages stress is different, and the mechanisms that you use to handle it can vary. In my view, the CIO and Risk Management Executive roles are quite alike in terms of the pressure that you can feel. The large difference is the degree of hands-on work that one has to do – what you can do to mitigate these risks versus providing challenge and guidance.

The role of the CISO has a greater level of inbuilt tension that must be dealt with daily. A CISO must react to scenarios that occur daily and often outside their direct control. There are "shadow boxing" exercises daily with false positives that must be treated as the real thing until proven otherwise.

There are some examples of the CISO moving into a CIO role; however, this is still relatively uncommon. Thus, Risk Management may be another strategic option to consider that builds a unique brand.

> **Key career tip**
>
> Your battle scars provide you with the perfect training to be a technology and cyber risk executive. What you have learned as a CIO and CISO in terms of SKEB can be easily transitioned from first-line roles into Risk Management.

As the profile of the Risk Management Executive changes, there will be a new set of responsibilities. It won't be good enough for the existing Risk Manager to not move with the new requirements. They will need to move to being more focused and helping the team win.

I do recall a lively debate that I had with some Risk Management Executive colleagues who insisted that their job was to write policy rather than to protect the enterprise. I was somewhat horrified, and visions of Risk Managers of years past came back for a few moments. What I said was: "No, our job is to protect the enterprise. How we do this is by risk oversight and insight. The policies exist to enable and support the risk appetite."

Risk Management

Worried about eliminating
large rocks to reduce real risk

---

risk management

More focused on small stones to
ensure process is working

Figure 11.1 – Capital R Risk Management versus risk management

I have always talked about *Capital R Risk Management* and *small r risk management*.

My job is to protect the company by getting the largest risks (the biggest rocks) removed. That to me is *Capital R Risk Management*.

My job also requires me to run Risk Management processes, and that is *small r risk management*. These are business-as-usual activities to make sure that the rules are being followed. The new era Risk Management Executive that I'm talking about is *Capital R Risk Management*.

In this section, we have explored the new requirements for Risk Management Executives and how this will drive an uplifted profile of the traditional Risk Manager. This will create a new career path for the CIO and CISO to seriously consider. Their technology skills, knowledge, experience, and behavior give them a strong advantage to meet the requirements of this new profile.

The next section I will introduce an analogy that will help you to understand this approach better. This uses the analogy of a football (soccer) game with the various characters from this scene.

Risk Management as coaching

Let's first understand the game that we are playing. In the company, we have the CIO and CISO, who are playing to win. This means delivering new value to the company and protecting the enterprise from cyber and technology disruption.

Figure 11.2 – CIO and CISO as the players

The CIO and the CISO are indeed in the game. They are *players* playing to win and have the right to make decisions at each moment in terms of where to pass and how to attack and defend. Their roles are hard and tiring, and they must take fast action to keep up with the crowd's demands. Should the team not win, then there is pressure from management to fire these players.

In the crowd, we have *spectators*. They are watching the game from a distance and cheering or booing. The spectators feel that they are playing the game but are not allowed on the field. They often will critique the game and provide feedback on what they don't like. The spectators are the various stakeholders in the company who are involved as supporters but are not actively in the game. In most enterprises, we have spectators that we must deal with as CIOs and CISOs.

Figure 11.3 – Stakeholders as the spectators

(Source: https://easy-peasy.ai/ai-image-generator/images/lively-latin-american-soccer-game-vibrant-stadium)

Then there is a *referee*. They are on the field, but they are there to interpret the rules and ensure that no one is cheating the system. In companies, we have many referees who are engaged in the game. They are often the Audit and Compliance team. The referee can penalize the players if they break the rules, so they should be respected and, at times, feared.

Figure 11.4 – Audit, Compliance, and Regulators as the referee

The referee is an unpopular role and sometimes must provide red or yellow cards for infringements. Regulators can often operate as a video referee and step in after awarding penalties if they feel this has been missed by the referee. The CIO and CISO will often feel that they are not treated fairly by all the referees and may want to dispute the facts.

Now, let's get to the heart of the matter. I have seen Risk Management Executives play the role of spectator, player, and referee. To my mind, this is why they are confused and don't really understand how they can help the team win.

Given that a new-age Risk Management Executive may have previously been a CIO or CISO, there is an alternative and more radical model, which is for the Risk Management Executive to act as the *coach*. The coach has experience as a player and understands all the tricks and traps of the game.

The coach is not on the field but sits high in the stands to watch the game and observe the patterns. While they will appreciate all the good things that are going on (like a spectator would), how the laws are being applied (like a referee would), and the effort on the field (like a player would), they are also able to provide an independent perspective that is different.

When the Risk Management Executive is a coach, they can suggest what the players are missing in terms of defense gaps or attacking opportunities. They bring their deep perspectives from having been a CIO or CISO to provide some alternative insights.

Figure 11.5 – Risk Management Executive as the coach

The coach may use data analytic metrics to talk about how the team is operating and how the opposite team is acting and reacting. A good coach wants the company to win. While they may not be able to play the game anymore, they have some insights from their own experience that provide wisdom that could be valuable.

This is my own experience. I have been able to move from a CIO role to being a CISO and then become a Risk Management executive. I've found the change to be rewarding, and this builds upon what I have previously learned, and I can apply those learnings in a different light.

As a coach, I *try not to overcoach*. What I mean by that is the most effective coaches don't try to fix everything at once. They realize that the players will get confused if they try to correct too many tactics.

The message must be simple and focused. For instance, what one or two things could you do differently to win? That requires real discipline from the coach, and they will draw upon their own battle scars and experience as a CIO and CISO.

By focusing on the key coaching message, the coach will help the team win.

In my case, I have been a coach of a basketball team and I've been a coach for aspiring CIOs. The thing that I learned on this journey is that what I say matters, but also what I don't say matters. To be a good coach comes with practice, so this is something you can try and test in a social arena before it becomes your job.

Let me provide a few examples. When you are the coach, you are mindful to try to limit your comments and coaching to what will win the game. Clearly, there are things you observe that aren't where they need to be, but every enterprise can only take on so many new things or projects. To win the game, you will need to be able to draw on your own experience and counsel the players to make changes to be more effective.

A very good example is to coach the players to focus on high-priority cyber controls that will have the greatest impact on the game outcome and are the most deficient. Conversely, a bad coach will ask the players to address all issues at the same time, forgetting that they are adding to the problem at hand with this approach rather than maximizing the risk buydown.

In this section, we have explored a football analogy and examined how the roles of the various characters are different in the new era of Risk Management. The former CIOs and CISOs can use their past battle scars and experience to be Risk Management Executives and provide specific technical coaching to help the team win.

The next section will provide some guidance and tips on becoming a Risk Management Executive.

Finding your way to become a coach

Now that you understand that this is a viable career option, we should explore how you get to be in the running for this role. In this section, we explore some approaches to making this shift. As we addressed in *Chapter 10, Looking for the Next Elevator*, we now have another alternate path to consider.

It is likely that the contacts that you have are not the ones who are commissioned to find Technology Risk Management Executives, but this is where I would advise you to start. Your relationships and their strengths are the places to start discussions. Taking the coffee catchup approach with some of your trusted HR search firms would subtly let them know that you are seeking a new CIO or CISO role or are thinking about an alternative path to becoming a Technology Risk Management Executive.

Once you have signaled that you are open to a potential move, then you can assess if your current contact can refer to you or still represent your interests. Otherwise, the process is the same as outlined in *Chapter 10*; there needs to be self-reflection and, in particular, thinking about how the role fits into your overall life career portfolio.

My advice is that taking the role of Risk Management Executive can be a real breath of fresh air. There is still pressure in the role because you still are there to protect the enterprise; it is just that you have a different perspective sitting in the coach's box compared to being on the field. I would highly recommend this path, and there have been many cases of Risk Management Executives returning to CIO and CISO positions.

At this point, I would like to provide another reason why you should consider Risk Management as a career path option. In my career portfolio, I have made it a mission to drive transformational change as a CIO and CISO. When I was considering my next career move from CISO at HSBC Asia Pacific, it was with the obvious two choices of CIO and CISO positions. However, my career plan was to reposition myself to be a Non-Executive Director and Board Advisor. This required some new SKEB. Therefore, I chose a Risk Management Executive career as a stepping stone to this career.

Summary

The road less traveled is a symbolic expression about choosing the less conventional or more challenging paths in life. Considering this option in addition to the dual paths of CIO and CISO may provide you with a richer career than you have considered at the start of this book.

There are many battle scars that CIOs and CISOs will have from working with Risk Management. The role of the Risk Manager is changing as requirements for more technology increase with the increased concern about cyberattacks and stronger focus by regulators.

A traditional Risk Manager has a different SKEB profile than the new-age technology Risk Management Executive.

On the playing field, the Risk Manager acts as the coach, not the player, spectator, or referee. The coach wants to use their own career experience to help the team win. This advice must be focused and not confuse the players.

The CIO and CISO should consider how their skills are complementary to the notion of a Risk Manager acting as a coach. This is a great career option, and in my next and final chapter, I will provide even more rationale for considering that path.

12
What CIOs and CISOs Do in Retirement

Congratulations! You scaled the mountain and reached the top. Indeed, you have summited numerous times and reached rarefied air and survived. Now what?

It is fascinating that there is an exclusive club of people who have summited all 14 of the highest peaks in the world. The members of this group have been dubbed the "eight-thousanders," as all 14 of these mountains are more than 8,000 meters high.

They include Mount Everest, K2, Kangchenjunga, Lhotse, Makalu, Cho Oyu, Manaslu, Dhaulagiri, Nanga Parbat, Annapurna I, Gasherbrum I, Broad Peak, Gasherbrum II, and Shishapangma.

I'm sure that most readers will only have heard of one or two of these mountains. The key point is that all of us must choose the mountain we want to scale. There may not be a gondola cab to give you an easier path up the trail; when we decide that we've climbed high enough, it is a key life/career decision point.

There will come a time when you will have had enough, and you will know that you want a change and don't find the CIO or CISO role as rewarding as you would like. The role will always be challenging – that's not what I mean. What you want will be different: your own motivation is massively individual and not the same as someone else's.

> The fundamental question that is to be addressed in this chapter is as follows:
>
> What do I plan to do in my retirement?

The decision to retire is not a simple one, as you will have been in busy and challenging roles. There are some people who, upon retirement, want to do very little and take life easy. Others who have been engaged and committed at work for a long time will then struggle to accept that their purpose in life is going to dramatically change.

This final chapter explores the career options that you can consider in retirement. For the aspiring CIO, CISO, or Risk Management Executive, this may appear to be looking too far in advance, but I always say that we must begin with the end in mind. Your retirement is a certainty that you must face. Hence, I suggest you skim-read this chapter now and then come back to it later in your career.

In this chapter, we'll be covering the following things:

- Looking at retirement as a new beginning
- Figuring out how old you should be when you retire
- Looking at a few post-career moves for CIOs and CISOs
- Planning your transition to boards
- Planning a transition into board advisory
- Climbing a different mountain

Looking at retirement as a new beginning

Retirement can be seen as a new beginning. Those who view situations as the glass being half-filled will see this as a step toward their death. There have been many studies about Bhutan being the happiest country in the world – do you know why?

There is within the culture of Bhutan a deep personal contemplation of death, and the average Bhutanese citizen thinks about death at least five times per day. Because they reflect on death so often, they choose to be happy. It is a form of mindfulness of living for the present. As you consider your own retirement, there will be many thoughts in your head. The first one is, are you ready, and why are you thinking about it? Some common reasons why the timing might be right for you are as follows:

Reason	Description
Financial security $\$$	You have reached a point where you don't need to work and have the personal freedom to choose not to. Your savings and investments will provide for you.
Work burnout	You have just had enough of the constant stress and are not feeling much motivation to carry on with the difficult parts. You have seen and experienced highs and lows in your career, and you just want something else.

Reason	Description
Health ⊕	As we age, there may be some health complications that make us all reconsider what is really important. You may have a wake-up call to take more care of your general health.
Lifestyle ✈	You might become motivated to follow other pursuits that are not work-related – travel, sports, family, etc. Lifestyle is a broad catch-all and likely to be applicable to many leaders.
Family 🧓	Taking care of aging parents and relatives is a common reason for wanting to retire. This reason is often more common in certain ethnic cultures where this is a common approach.

Table 12.1 – Retirement key considerations

These five considerations should all be taken into your thought process. None of these in particular is objectively more important than the other; in your post-work life, you will want to balance all of these.

Every leader will want to figure out how to live a healthy life and enjoy their golden years. At the same time, you will have a voice in the back of your head asking, "Am I still on top of my game?" Clearly, no leader wants to be hanging onto their role and feeling that their best days are behind them. All of us would like to retire on our own terms and set our own timeline rather than having someone else do it for us.

In the next section, we explore a simple but hard question –how old should you be when you retire? There is no right or wrong answer; rather, it depends.

Figuring out how old you should be when you retire

This is a very personal question, and it depends upon how you respond to the factors in the preceding table. There is no objectively wrong or right time. I used to play golf with some retirees when I lived overseas and came back to Sydney for my holidays. The golf players on a weekday would be anywhere between 60 and 80 years of age, and I would meet many of them during these games. My questions were always intended to understand their perspective:

- When did they retire?
- Was it too early or too late?
- What would they do differently?

There were no real patterns or insights that I could gather. It just struck me that this was a deeply personal decision that had to be made with solid reflection. Therefore, planning for retirement is as important as planning for your career.

I learned on the way that there are some other questions beyond "how old should I be to retire?" While I don't have a template for retirement similar to our previous 90-day plan, here are a few of my thoughts that I have considered in my own plan:

Retirement planning	Considerations
Retirement goals	<ul><li>When do I want to retire?</li><li>How will I spend my time?</li><li>How will I keep myself mentally active?</li></ul>
Solid financial planning	<ul><li>Talk to a financial advisor</li><li>Review my financial plan</li></ul>
Location	<ul><li>Consider where I want to live</li><li>Research options</li><li>Do a test run of new locations</li></ul>
Retirement projects	<ul><li>Develop a list of personal projects that I want to tackle</li><li>Consider some not-for-profit and volunteering work</li></ul>
Professional project	<ul><li>Build a board CV</li><li>Develop a board network</li></ul>

Table 12.2 – Planning for your retirement

There is much reflection required, and some of these activities should be done with your partner rather than individually. Once you have completed your own retirement planning assessment, you can consider some of the options available to keep your brain active. The next section explores post-career moves that the CIO or CISO can consider.

Looking at a few post-career moves for CIOs and CISOs

Your career as a CIO/CISO will have provided you with a great network and a set of very useful skills for your post-career. There are a series of logical paths that you can take, which include these:

- Non-executive director board roles
- Board advisory consulting

- Strategic consulting

- Start-up advisor

- Keynote and conference speaker

- Authoring and writing

For those that want to pursue a portfolio career of a selection of these, then a career awaits that is going to be real fun! Each of these activities can be fun and provide you with further learning during your retirement. To me, the exciting part is combining these elements into your post-retirement plan.

The typical CIO or CISO will find it difficult to stop all work and cease activities. You will have been wired each day with constant challenging stimuli and problems to solve. Each of the following activities can provide you with some mental stimulation:

Professional role	Commentary
Non-executive director board	A CIO or CISO will have spent much of their career answering questions from the board. It is logical that they may want to take on similar positions and guide other organizations. There is an increasing demand for tech-savvy board members who can help guide companies going through digital transformation and develop cyber resilience.
Board advisory consulting	There are board advisory roles that can be taken up via boutique consulting firms or directly, where a board risk committee appoints a board advisor with a tech background. The board advisor is able to provide guidance and is not normally subject to the director's liability.
Strategic consulting	An experienced CIO or CISO can help other C-suite industry colleagues with strategic guidance and support. They can provide input for key decisions or new strategic plans.
Start-up advisor	There are many start-ups that are scaling up and don't understand how to position their products to CIOs and CISOs. They can also benefit from the strategic thinking that retired CIOs and CISOs can provide. These roles can be particularly rewarding, as they are often more hands-on.

Professional role	Commentary
Keynote and conference speaker	The retired CIO/CISO has many insightful *war stories* that they can share with their peers and industry colleagues.
	Keynotes, panels, and interviews can be rewarding; it can be especially challenging to manage different interview questions tactfully.
Authoring and writing	While not for everyone, this is the next step from keynote conference sessions, and committing to sharing your stories more formally can be stimulating.
	Writing is both a left- and right-brained activity, so it's great mental exercise for the retiree.

Table 12.3 – Post-career options

You have to judge for yourself whether you want to try all of these options or just focus on a few. For some people, none of this will feature in their plans and instead they will just want to focus on charity work and giving back. There is no right or wrong.

In the next section, I will provide a few more thoughts about the first two post-retirement roles discussed in the table. These are classically what CIOs and CISOs will take on in their retirement. The other options are less mainstream and not for everyone.

Planning your transition to boards

As a CIO or CISO, a new way of thinking is required for board work. You must understand how these roles are subtly different from what you have been doing and make an adjustment. What is valued in a CIO or CISO will remain the same, but as a board member, you must operate at a higher strategic level. Here are a few things you should do:

- **Develop your plan at least a year out from considering retirement**: It would be advisable to start to shift your mindset through non-executive board training and completion of the certification process. There are various institutions around the world that provide this training and I'd highly recommend this path.

- **Evaluate your value proposition as a board member**: What do you bring that is valuable? Boards will often want to leverage recent management experience from people who have dealt with challenges such are remote work during COVID, how to manage generative AI, and so on.

- As a CIO or CISO, there will often not be many people who have your background on a board. If this is the case, then it is likely that you might be seen as the de facto technology expert. While this may be true, you will want your board colleagues to not just defer to you on all tech and cyber questions, as this would perhaps not be the best outcome in the long term.

- **Define what you see as the best fit for your board role**: Decide on industry, company, location, size, and what the overall company mission is. If you have good alignment with the mission, that is a great start. Evaluate what digital transformation or cyber/data experience would be valued by a specific industry or player.

- **Rewrite your CV from scratch**: A new board CV is usually no more than two pages, so you will need to summarize the long-form profile of 5-8 pages and only cover what would be relevant for a board position.

- **Build your board network**: Build a network with search firms and current board members. This is a vital step, and without effectively doing it, your opportunities will be limited. The search firms are not dissimilar to the executive firms that you will have dealt with in the past; the only key difference being that there might often be a distinct team for board practice. Thus, you must get referrals to meet the respective gatekeepers. It is true that your network is a major source of recommendations and referrals for new board roles. Therefore, strategic networking can really help your search.

- **Research target companies**: This will mean reviewing the current board members – who is about to retire and who knows who? That personal network can help you talk and meet with current board members or, even better, the chair of that board.

In this section, we explored the steps you should take to move into board roles. All the skills, knowledge, experience, and behaviors you will learn on your CIO/CISO journey are great foundations for this new career. But as with all transitions, it will require some reflection and adjustment.

In the next section, we examine board advisory roles.

Planning a transition into board advisory

Board advisory roles can be a means to an end, in that they can lead to a full board role, or they can make for rewarding post-CIO/CISO careers in themselves. When you are being considered for a board advisory role, know that it will be a subject matter expert position for the full board or a sub-board such as a risk committee or technology committee.

As an advisory board member, you do not have any voting rights. Your role exists to provide expertise and help guide the company through some strategic change or transformation. What you are valued for would include the following:

- Independent strategic advice
- A sounding board for key decisions

- Mentorship for existing board members

- Assisting the company with insights and perspectives

Board advisory is a common pathway for the CIO or CISO to take. There are many formal and informal alternatives out there. Here are a few ideas to try:

- **Scan the market for board advisory roles**: These roles are not always advertised and can come from many different sources. You may have prior board members as contacts, and they can be leads. The required commitment might be short-term to provide technology expertise to a risk committee; this is often required for a large transformation or cyber uplift. The time commitment could even be one day, a month, or a fortnight. It can vary greatly, and you might be engaged for years in this capacity.

- **Tailor your advisory CV**: Again, your CV must be tweaked to be relevant to the board advisory roles you seek. Providing a stronger emphasis on recent experience with different subject matter areas that you can provide expertise on should be the focus. You are a hired gun, so this is not a long-term job; hence what matters is what expertise you can immediately bring. It may be a strategic project or a recent transformation/remediation effort.

- **Set up a company structure**: You will need to do some setup of how your board advisory will be invoiced and the admin structure around this. Therefore, you may need to talk to your accountant and/or tax advisor for some simple advice.

In this section, we have discussed taking on board advisory work. Advisory roles can be short and sharp but very rewarding when you deliver a specific technology, data, or cyber outcome. The work can feel quite familiar to you being back in an office; however, there is a defined endpoint, and this can be quite motivating. There is a subtle difference to strategic consulting in many regards; the process is the same. The key differences are in who your key customer is and that strategic consulting can be 1–5 days per week.

Maybe none of the options I outlined here interest you; this is a personal decision. In the next section, I touch on scenarios where you don't want to do this anymore and have a different mountain in mind.

Climbing a different mountain

For some CIOs and CISOs, they don't want to see another board room ever again, and their retirement is all about grandkids, golf, travel, or sitting on a beach.

Each of these is also a great option. It just depends on how you are wired and whether you can adjust from your career to the post-career stage and be happy with this change.

Good luck with retirement – finding true purpose beyond work is important. Many will have spent over 40 years working in different roles to reach this point and can choose what to do next.

Retirement to me is all about focusing on living well and building your bucket list. Your career as a CIO or CISO will have been a long journey of learning, and this stage is no different.

Summary

You have climbed the mountain and must contemplate what is next. Retirement and not climbing any more mountains is an option we all face now. Taking a proactive approach to planning your retirement and how you will retire is just like planning other stages of your career.

There are career options available to the retired CIO or CISO, such as board roles and board advisory, and other alternatives such as consulting. These roles can use your experience; however, you will still have to consider and learn how to change your approach, as this requires some subtle change.

Planning the transition is key, and you must adapt as you will have done throughout your career. Paul McCartney of the Beatles wrote *The Long and Winding Road* as he looked upon a road stretching into the hills in Scotland. Your life and your career are symbolized by the twists and turns that you take up your mountain.

Enjoy the journey.

Appendix

Here are some key questions to ask yourself as you read this book. This is an optional exercise, but I would suggest that you attempt to answer these after you finish reading the book.

My hope is that you have greater clarity by that point. This can also serve as a review to allow you to further reflect before starting your journey.

1. Why do I want to be a CIO? (What are the reasons?)
2. Why do I want to be a CISO? (What are the reasons?)
3. Do I really understand the roles of a CIO and CISO?
4. Do I understand how to become a CIO and/or CISO?
5. Should I consider both roles in my career plan?
6. What is my current personal brand? (What do others say about you?)
7. Do I know what my own gaps are to be a CIO and/or CISO?
8. What are my specific skills gaps?
9. What are my specific knowledge gaps?
10. What are my specific experience gaps?
11. What are my specific behavior gaps?
12. What are the top 10 skills of a CIO and/or CISO?
13. What soft skills do I need to develop?
14. Which of my skills is my superpower that I should draw on?
15. Should I have a career objective in my CV?
16. Should I have a position objective in my CV?
17. How do I plan to be comfortable with being uncomfortable?
18. Do I understand my own IQ, EQ, AQ, and SQ gaps?
19. How do I think two jobs ahead?
20. What is the algorithm to accelerate my career growth?
21. How does developing my team help me grow and develop?
22. Do I understand the traditional career path to be a CIO and/or CISO?
23. Do I understand the linear career path to be a CIO and/or CISO?

24. Do I understand the transitory career path to be a CIO and/or CISO?

25. Do I understand the spiral career path to be a CIO and/or CISO?

26. How do I find this new position as CIO and/or CISO?

27. How should I prepare for the CIO and/or CISO interview?

28. What CIO or CISO interview questions could I be asked?

29. What questions should I ask at my CIO and/or CISO interview?

30. How do I write my CIO and/or CISO first 90 days plan?

31. How do I engage critical stakeholders?

32. How do I run my first meetings with my stakeholders?

33. Why should I do stakeholder analysis?

34. How do I build an influence map?

35. How do I learn about hidden agendas?

36. How do I assess my new team?

37. How do I build a future team?

38. How does my new team engage and why is this important?

39. How do I get the team to respect what I inspect?

40. What CIO and CISO personal key metrics do I need?

41. How do I accelerate my business learning?

42. How do I fix critical *hygiene* issues?

43. How do I challenge and revise my IT strategy?

44. How do I challenge and revise my cybersecurity strategy?

45. How do I understand my IT and cyber operations?

46. How do I improve my team's culture?

47. Do I have the right cyber governance approach?

48. What is my security baseline?

49. How do I manage my regulatory book of work?

50. What are the pressures that a CIO and CISO face?

51. What can cause tension in the relationship between a CIO and CISO?

52. How do I manage the stress that comes with this CIO and CISO role?

53. What is a transformational CIO and how is their stress different?

54. What is a strategic CIO and how is their stress different?

55. What is an operational CIO and how is their stress different?

56. What is a partnering CIO and how is their stress different?

57. What is an empowering CIO and how is their stress different?

58. What is a technical CISO and how is their stress different?

59. What is a transformational CISO and how is their stress different?

60. What is an operational CISO and how is their stress different?

61. What is a business CISO and how is their stress different?

62. What is a partnering CISO and how is their stress different?

63. What is a risk culture CISO and how is their stress different?

64. Will I feel the weight of being a leader?

65. How do I understand the expectations of the board?

66. How do moments of truth accelerate my development?

67. What are my survival tips for being a successful CIO and CISO?

68. How do I manage political situations?

69. What does Maslow's theory mean for a CIO and/or CISO?

70. How do I nail the interview for my next CIO and/or CISO role?

71. How should I weigh the challenge, compensation, and industry/location for any new role?

72. How do I understand the power and interest of different stakeholders?

73. What are the criteria I should use to select my next CIO or CISO role?

74. Can the CISO trust the CIO (and vice versa)?

75. How do I build this trusted relationship?

76. Who are my alliance partners (friends and foes)?

77. How can I use Myers-Briggs to assess my stakeholders?

78. How are the CIO and CISO aligned?

79. Should I consider getting a mentor?

80. Should I consider getting a coach?

81. How should I look for my next elevator?

82. How do I make significant career and life decisions?

83. How do I use the transit lounge to position my next role?

84. Do I understand how a career in consulting works?

85. How do I build my career through a portfolio of experiences?

86. Should I consider moving from the CIO to the CISO role?

87. Should I consider moving from the CISO to the CIO role?

88. Should I consider an international assignment?

89. Should I consider changing industries and sectors?

90. How do I plan for my handover to my successor?

91. Why should I care about doing a great handover to my successor?

92. How do I balance my career ambitions and my family needs?

93. How do I prepare for life after being a CIO and CISO?

94. Should I consider risk management as a career path?

95. What skills, knowledge, experience, and behavior do I bring to a risk management career?

96. Can I learn to ask better questions by being in risk management?

97. How does a risk management career help me with being on boards?

98. When should a CIO and CISO consider retirement?

99. When I plan for my retirement, what should I position myself to do?

100. What skills, knowledge, experience, and behavior will I need for this post-retirement career?

I anticipate that you have reflected on these questions and have attempted to address each of them. You should now be well equipped to manage your career from wherever you are starting to achieve your personal goals.

Index

‹packt›

packtpub.com

Subscribe to our online digital library for full access to over 7,000 books and videos, as well as industry leading tools to help you plan your personal development and advance your career. For more information, please visit our website.

Why subscribe?

- Spend less time learning and more time coding with practical eBooks and Videos from over 4,000 industry professionals
- Improve your learning with Skill Plans built especially for you
- Get a free eBook or video every month
- Fully searchable for easy access to vital information
- Copy and paste, print, and bookmark content

Did you know that Packt offers eBook versions of every book published, with PDF and ePub files available? You can upgrade to the eBook version at packtpub.com and as a print book customer, you are entitled to a discount on the eBook copy. Get in touch with us at customercare@packtpub.com for more details.

At www.packtpub.com, you can also read a collection of free technical articles, sign up for a range of free newsletters, and receive exclusive discounts and offers on Packt books and eBooks.

Other Books You May Enjoy

If you enjoyed this book, you may be interested in these other books by Packt:

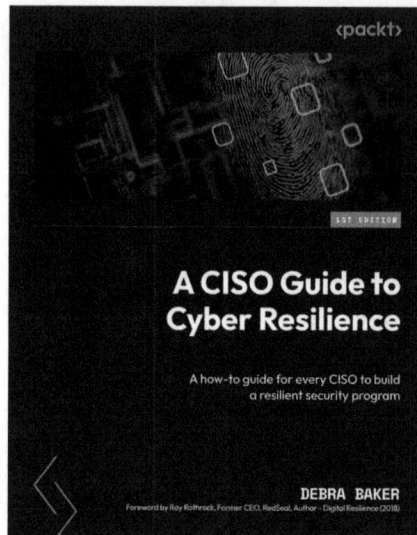

A CISO Guide to Cyber Resilience

Debra Baker

ISBN: 978-1-83546-692-6

- Defend against cybersecurity attacks and expedite the recovery process
- Protect your network from ransomware and phishing
- Understand products required to lower cyber risk
- Establish and maintain vital offline backups for ransomware recovery
- Understand the importance of regular patching and vulnerability prioritization
- Set up security awareness training
- Create and integrate security policies into organizational processes

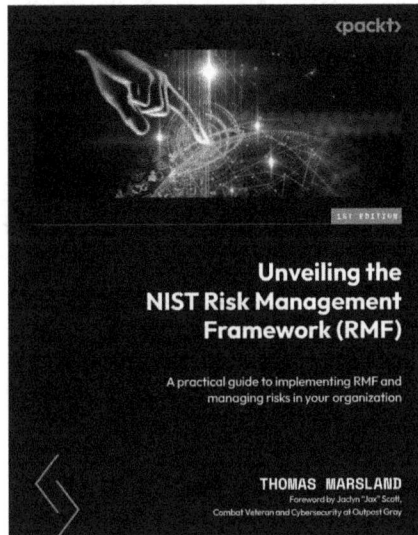

Unveiling the NIST Risk Management Framework (RMF)

Thomas Marsland

ISBN: 978-1-83508-984-2

- Understand how to tailor the NIST Risk Management Framework to your organization's needs
- Come to grips with security controls and assessment procedures to maintain a robust security posture
- Explore cloud security with real-world examples to enhance detection and response capabilities
- Master compliance requirements and best practices with relevant regulations and industry standards
- Explore risk management strategies to prioritize security investments and resource allocation
- Develop robust incident response plans and analyze security incidents efficiently

Packt is searching for authors like you

If you're interested in becoming an author for Packt, please visit `authors.packtpub.com` and apply today. We have worked with thousands of developers and tech professionals, just like you, to help them share their insight with the global tech community. You can make a general application, apply for a specific hot topic that we are recruiting an author for, or submit your own idea.

Share Your Thoughts

Now you've finished *The Aspiring CIO and CISO*, we'd love to hear your thoughts! Scan the QR code below to go straight to the Amazon review page for this book and share your feedback or leave a review on the site that you purchased it from.

`https://packt.link/r/1835469191`

Your review is important to us and the tech community and will help us make sure we're delivering excellent quality content.

Download a free PDF copy of this book

Thanks for purchasing this book!

Do you like to read on the go but are unable to carry your print books everywhere?

Is your eBook purchase not compatible with the device of your choice?

Don't worry, now with every Packt book you get a DRM-free PDF version of that book at no cost.

Read anywhere, any place, on any device. Search, copy, and paste code from your favorite technical books directly into your application.

The perks don't stop there, you can get exclusive access to discounts, newsletters, and great free content in your inbox daily

Follow these simple steps to get the benefits:

1. Scan the QR code or visit the link below

https://packt.link/free-ebook/9781835469194

2. Submit your proof of purchase
3. That's it! We'll send your free PDF and other benefits to your email directly